BASIC AND INTERMEDIATE CELESTIAL NAVIGATION

BASIC AND INTERMEDIATE CELESTIAL NAVIGATION

WM. BRUCE PAULK

HEARST MARINE BOOKS NEW YORK

Library of Congress Cataloging-in-Publication Data

Paulk, Wm. Bruce.
Basic and intermediate celestial navigation / Wm. Bruce Paulk.
p. cm.
Bibliography: p.
ISBN 0-688-08939-9
1. Navigation. I. Title.
VK555.P34 1989
623.89—dc 19 89-30771
CIP

Printed in the United States of America

First Edition

1 2 3 4 5 6 7 8 9 10

BOOK DESIGN BY JAYE ZIMET

TO ENID,
MY FIRSTMATE AND SOULMATE

LAHAINA, MAUI, HAWAII

PREFACE

This is not just another book on celestial navigation. Before I go into detail on the approach and contents, I feel the need to tell you a true story that deeply concerns me.

It was our second evening in Lahaina. My wife Enid and I had just completed a 24-day passage from Morro Bay, California to Lahaina, Maui, Hawaii in ZOE, our 35 ft. Ericson sloop, and had decided to dine at the yacht club that evening. While enjoying our dinner we were approached by a man who engaged us in conversation. He stated he had seen our boat in the harbor and had heard of our passage from the mainland. He told us he had arrived in Lahaina after a very long passage from the coast several days prior to us and began to elaborate on certain aspects of his passage that were frightening.

It seems that he and his crew had set sail for the Hawaiian Islands with a brand new electronic navigation device. Neither he nor any member of his crew had any previous celestial navigation experience, and they didn't even have a sextant aboard. They did have a chart of the North Pacific Ocean and were plotting their positions as obtained from periodic satellite updates enroute.

About halfway across, the device ceased to operate, and they had no other means of obtaining terrestrial positions available to them. The morale aboard immediately began to deteriorate, arguments ensued, and whoever was on watch steered a course to his own liking, all in hopes of finding a ship and being rescued. Eventually, someone noticed the contrails from jet aircraft heading in the general direction of the Hawaiian Islands and they decided to follow the contrails. As fate would have it, they sighted an island, and after using the radio telephone for instructions from other mariners, were told they had found Maui. They were advised to sail around to the leeward side of the island, and they eventually found their way into Lahaina.

I believe that the crew of that yacht was extremely lucky to have completed its passage. The contrails cannot be trusted because many routes appear to be going to the islands but are not. Also, they can't be seen a large percentage of the time. The islands could easily have been missed completely. They were actually in greater danger than the early explorers. The explorers had the ability to keep a watch from aloft in the crow's nest, usually had enough seagoing experience to know when they were approaching land, and could determine their latitude with an early version of the sextant. I wouldn't want to make a passage without daily knowledge of my position, especially when making a landfall. The chances of approaching the unlit windward side of an island in darkness are about 50%, and often visibility is obscured by cloud cover and haze.

It is my personal belief that the art of celestial navigation should be learned by anyone who might possibly venture out of sight of land. This category includes coastal cruising sailors. Many of these sailors lose their bearings in the fog and occasionally have to heave to on a seaward tack in untenable weather conditions. When the fog or weather improves, celestial navigation could be utilized to establish a terrestrial position and a harbor could then be approached with reasonable safety.

Even in sight of land, without adequate landmarks, one's position along a stretch of coastline can be deceiving. In this instance, a noon sight establishing the latitude would be comforting.

This all brings me to a favorite quotation from *Dutton's*,

> "At sea, the proper practice of celestial navigation is of the utmost importance. Even in areas where electronic navigation systems provide good coverage, celestial navigation must not be neglected: Electronic systems or individual pieces of equipment may suddenly become unavailable. The prudent navigator uses every available means of fixing his position."

I am 100% in favor of using electronic navigation devices like Sat. Nav. and Loran C. There is no doubt that electronic devices can do a faster and more precise job of locating a terrestrial position at sea, if they are operating, and more important, if the navigator knows how to utilize them properly.

My well-founded main concern is for you to first learn celestial navigation, and practice it whether these electronic devices are installed or not. Celestial navigation knowledge and experience is the backup, not electronic navigation devices. An electronic device is only one of the tools available to the navigator.

And that is why I have written this book which is really two books in one. The first book is a complete text on basic celestial navigation utilizing the most compact, concise tabular method of sight reduction available to the small craft navigator. The tabular method of sight reduction you will be using requires no special mathematical skills. Anyone who can add and subtract will be able to use these tables. I utilized a scientific calculator for the voyage described in the text. However, this voyage is the basis for all the navigational problems and the tables necessary to solve them have been incorporated into the text so that you can work out the answers along with me. An outstanding feature of the sight reduction tables you will be using is that they exceed the accuracy required for small craft navigation.

In the text, you will be sailing with us on our passage from California to Hawaii during September of 1986. You will be sharing many of our experiences, and in addition to finding our position at sea, you will be planning our passage, plotting our daily dead reckoning position, determining our daily terrestrial position, updating our course, and sharing our landfall off the island of Maui. The tools required for practicing celestial navigation are listed and discussed. Complete, step by step instructions are given for adjusting your sextant and taking sights of the navigational celestial bodies. You will be sighting these celestial bodies enroute. The text contains complete information required for observing the sun, moon, stars, and planets.

Once the basic material is understood and practiced, you could navigate your own yacht anywhere in the world with confidence. If it is your desire, however, to become a better, more competent navigator, Book Two provides the basis for this transition.

Book Two is an intermediate text on celestial navigation and begins with the basic theory, including a comprehensive but understandable treatment of the navigational triangle. Knowledge of the navigational triangle is necessary to complete the final portion of the course.

Book Two contains complete instruction in celestial navigation using a scientific calculator and focuses on solutions to the navigational triangle. The calculator is utilized for sight reduction, calculating latitude, longitude, sunrise, sunset, moonrise, moonset, civil and nautical twilight, dip, and refraction. Use of the calculator in passage planning, complete with great circle and mercator sailings is covered in detail.

Before attempting this course in celestial navigation, it is recommended that you be familiar with the principles of coastal navigation referred to as piloting. Courses in piloting are taught by the U.S. Coast Guard Auxiliary and the U.S. Power Squadrons. Excellent text material is available in the classic book: *Chapman Piloting, Seamanship, and Small Boat Handling*, by Elbert S. Maloney.

I wish to acknowledge my gratitude to my wife Enid for drafting the illustrations and donating the sketches in the book. I also want to thank her for patiently handling other details of our daily life so that I could prepare this material for publishing.

CONTENTS

BOOK 1
BASIC CELESTIAL BY TABULAR METHODS

BOOK 2
INTERMEDIATE CELESTIAL BY COMPUTING METHODS

INTRODUCTION

There is a great sense of personal achievement in celestial navigation. When you have found your way across the trackless sea, without the aid of outside systems and electronic equipment, *you* have done the navigating on your own. You will, of course, be planning and plotting your courses, but if you use positioning data from radio-navigation systems, you have had "outside help" and can't claim full credit for the navigation, no matter how accurately you close on your destination. Out of sight of land, and with no lights or buoys, you are working with only your sextant, time information from some source, and a set or two of data tables. *You* are dependent on *your* personal knowledge and skills for your safe passage and arrival. And when you have done it successfully, the feeling of accomplishment can be compared to few other actions—you are a "navigator" in the truest sense!

Celestial navigation is probably navigation in the purest sense of the word, and it has a proud history of several centuries. Almanacs listing the stars' positions were among the first books ever published; sextants have evolved over the years; and instruments for obtaining and maintaining accurate time have advanced from the mechanical to the electronic. Changes have sometimes been slow in coming, but they have always been toward easier procedures and more accurate results.

Methods of sight reduction also developed over the past three centuries. Many sets of equations were used, progressing from simple listings of trigonometric values to highly complex tables of precalculated solutions and the so-called "inspection tables." Hydrographic Office Publications 214, 229, and 249 are well-known names to navigators, but each came in sets of multiple volumes, often to the serious inconvenience of the small-craft skipper/navigator.

Several more compact tables have appeared in recent years—the most recent being the introduction of the "Davies" tables. These were first available as a commercially published individual volume, but are now a part of each year's *Nautical Almanac.* This new sight-reduction method does not offer any increase in the precision or accuracy of the procedures, but it does offer a considerable convenience as now only a single volume is required, rather than an almanac for astronomical data and a separate set of tables for calculations.

Although much off-shore cruising is now done using radio-navigation systems, every skipper must always keep in mind the fallibility of electronic equipment. Each of these systems requires complex equipment both on the vessel and on shore—troubles at either end can mean at least a temporary loss of navigational information, and equipment failure on board can mean a loss of the system until a port is reached that has the requisite technical support facilities. It is at such times that celestial navigation becomes essential, and the wise navigator practices it even when other methods are being routinely employed in order that he may be knowledgeable and experienced when the need arises. This book records the experiences of a typical ocean passage using celestial navigation. It serves both as an excellent guide to learning the latest method, and as an interesting description of life at sea in a small sailboat.

—Elbert S. Maloney

BOOK 1

BASIC CELESTIAL BY TABULAR METHODS

ONE

GETTING STARTED

How did I become a cruising sailor and acquire the necessary skills in *celestial navigation*? We all have to start someplace. When I was in high school I was invited to go sailing one afternoon in a friend's Snipe-class sloop. Although I truly enjoyed myself, it was many years before another opportunity to sail presented itself.

Later, while attending the University of California at Berkeley one of my physics instructors was looking for someone to "do a little varnishing" on his 38-foot wooden sloop. I needed some extra money and volunteered. The sloop was berthed in Sausalito, across the bay from San Francisco. I enjoyed working on the yacht, and it wasn't long before I was asked to go sailing. This time the addiction took hold, and it has never released me.

At college I had many opportunities to crew on yachts entered in the Yacht Racing Association regattas on San Francisco Bay. I took a course in sailing given by the university, and was asked to assist the instructor for the last two semesters while I was in attendance. It was wonderful. I was getting paid for what I wanted to do. I even built a Pelican-class sloop and became a half owner in a Mercury-class sloop before graduating from college.

After obtaining my A.B. degree in physics, I went to work and pursued the acquisition of my first yacht in earnest. I bought a new 26-foot Ariel-class sloop and cruised and raced her for over five years on San Francisco Bay. I occasionally made short coastal cruises and started to single-hand the 26-foot ZOE around the Farallone Islands and to other exotic places like Drakes Bay, Half Moon Bay, Santa Cruz, and the California delta.

It wasn't long before celestial navigation became a necessity. The only course in celestial navigation I knew of utilized *H. O. 211* or *Agetons method.* A tremendous amount of time was spent in a planetarium discussing the position of the constellations and I barely learned to figure a *line of position* before the class was over. I knew there had to be better information available. I was fortunate to have a strong physics and mathematics background and could put together the basics from the major texts on this subject, e.g., *Bowditch* and *Dutton's*.

In the meantime, I traded in my Ariel for a new Ericson 35-foot sloop.

Two months later I received a promotion and was transferred to Honolulu in the early part of 1971. One of the conditions of my relocation was that my employer would ship my new sloop to Honolulu. The new ZOE couldn't be sailed to the Hawaiian Islands because she was not yet equipped for a passage of this scope and was barely in sailaway condition.

Relocating to the Hawaiian Islands was the best thing that could have happened. I met many cruising sailors who were actually making long passages and freely shared their navigational experiences. Soon I became aware of the finest tabular method of sight reduction available to small boat sailors at that time, referred to as *H. O. 249*. The three volumes of sight reduction tables formerly called *H. O. Pub. No. 249, Sight Reduction Tables for Air Navigation* have been renamed *Pub. No. 249, Sight Reduction Tables for Air Navigation*.

While sailing inter-island I practiced celestial navigation and within a short period had made my first crossings to and from California. The practice of celestial navigation became increasingly interesting to me and I began to document my experiences which became the basis for this book. In the book you will be learning celestial navigation while sailing from California to Hawaii with my wife and me. I believe sharing our voyage with you will help you to get a better feeling for celestial navigation and bring its practice to life.

In order to proceed with Celestial Navigation, you will require some basic tools and materials. The following is a shopping list and discussion of the selected items:

1. Sextant—any nautical type.
2. *The Nautical Almanac* for the current year.
3. Pad of *Universal Plotting Sheets No. VP-OS*.
4. Dividers—nautical or commercial type.
5. One of the following: Parallel rulers, parallel plotter, plotter, protractor, 30°/60° and 45°/45° triangles.
6. Pencils, erasers, and manual pencil sharpener.
7. Quartz-crystal wristwatch or alarm clock.
8. Shortwave receiver for Stations WWV and WWVH on 2.5, 5.0, 0.0, and 15.0 MHz, or a Time Cube or suitable equivalent.
9. One or two stopwatches.
10. Supply of Line of Position Worksheets.
11. Star Finder No. 2101-D.
12. Scientific calculator.

The sextant can be any nautical type. Do not purchase an old bubble sextant. These instruments were designed for air navigation.

If you don't live near the water you will require an *artificial horizon* to be used in conjunction with your nautical sextant.

I personally recommend a plastic or light metal-type sextant for small-craft navigation. The accuracy of these instruments is well within allowable limits for observations from a small boat. The lighter weight assures ease of handling, especially on the sea where there is frequently a lot of severe motion. Think twice before buying one of the more expensive, heavy, brass instruments which are for sale. They can work against the yacht navigator who is trying to maintain balance in a rolling sea, and hold the instrument still as well.

I recommend the sextant be fitted with a *micrometer drum* rather than a *vernier scale*, although the vernier type sextant is the least expensive. On a micrometer drum sextant, the drum is turned for the fine-tuning of the sight, and the reading of degrees and fractions thereof are taken from a scale on the drum. The vernier sextant has a vernier scale at the bottom of the index arm which is read against the degree scale on the sextant arc.

Keep in mind that when taking sights from the stable bridge of a large ship it is possible to achieve accuracy of less than a nautical mile. On a small vessel, while obtaining sights in heavy seas, accuracy of five nautical miles is more likely. Be aware of this limitation when approaching a landfall or a course hazard.

I have used two very good plastic sextants, with micrometer drums, on my passages . . . an English Ebbco and a Davis Mark 15. I also carry a vernier type Davis Mark 3 in my liferaft survival kit. The only addition I have considered is a lightweight aluminum alloy sextant with greater magnification for star observations. The approximate costs of marine sextants are as follows:

Lifeboat plastic sextants . . . . approximately $25 to $50
Micrometer drum plastic sextants. approximately $50 to $175
Light metal sextants . . . . . Approximately $350 to $600
Top line brass sextants. . . . approximately $1200 to $1700

You will also need a copy of the *Nautical Almanac* for the current year. The *Nautical Almanac* is published by the U. S. Government Printing Office in Washington, D. C. 20402, and Her Majesty's Stationery Office at 49 Holborn, London w. c. 1.; and there are several yachtsman editions available. If you should obtain a yachtsman edition of the *Nautical Almanac*, make certain that it contains all the information printed in the government publications.

The *Nautical Almanac* contains all the information you will require for the completion of this course. In the few instances where alternative methods are discussed the information is provided in the text.

You will require a pad of *Universal Plotting Sheets No. VP-OS* for plotting *dead reckoning* positions and lines of positions. The plots will show your most probable position on the earth's surface, and are called a *simultaneous fix* or a *running fix*.

Dividers will be used to measure and transfer distances and positions. I prefer the one-handed nautical type but draftsman's type dividers are also acceptable. Make sure that the dividers are at least six inches long so that you will be able to walk distances across a plotting sheet or chart with ease.

Many paragraphs could be written regarding plotting tools. I will briefly cover this subject. Your own experiences with these tools will in time determine your preference. The traditional tool for walking bearings across plotting sheets and charts is the parallel ruler. I use this sometimes, but I prefer the parallel plotter that is constructed with an axle fixed to two small metal, non-skid, wheels on either end. With this device you can roll your bearing across the sheet. A drafting type is acceptable or a nautical type such as a Weems Parallel Plotter can be purchased.

Another type of plotter used for both nautical and air navigation combines a 180° protractor and a straight edge. Make sure that the graduations on the scale on the straight edge are in increments of twenty nautical miles to the inch. This is the same scale that is used on the plotting sheets.

Another useful tool is a circular protractor. I have found the circular protractor especially useful when plotting bearings on a chart.

I've also found it helpful to use two drafting type plastic triangles, six inch types, one 30°/60° and one 45°/45°. They are excellent for drawing perpendicular lines and two triangles can be used to transfer a parallel line.

All of the above plotting tools are constructed of plastic and are relatively inexpensive.

You may wish to purchase a dozen or so wood pencils, a couple of soft erasers, and a hand-held manual pencil sharpener. The pencil sharpener should be the small hand-held type, about an inch long, that is rotated.

Lately, I have been using a mechanical pencil of the type where the top is depressed to advance the lead. The proper size of the lines you will be drawing is 0.5 mm, and a pencil to accommodate leads of this dimension should be purchased. I recommend purchasing several of these pencils and an adequate supply of leads.

Do not, under any circumstances, waste your money on a manually-wound *chronometer*. Today's inexpensive quartz-crystal wristwatches and alarm clocks keep better time and have a more predictable, discernable error.

Recently my wife gave me a Casio Alarm Chrono wristwatch

that I used on our last passage from Morro Bay to Lahaina. It is a quartz-crystal watch with digital readout for day of week, month, day, and time to the nearest second. The watch has many usable features such as a stopwatch, lap time, etc. The watch is suitable for navigation and gained only two seconds per week enroute. I deducted the two second error as our voyage progressed. Deducting the error rather than attempting to reset the watch while enroute is the preferred practice. I continued to monitor the error by listening to the time announcements on Stations WWV and WWVH while listening to marine weather broadcasts.

My wristwatch was backed up by two quartz-crystal alarm clocks with fresh batteries. These clocks were also checked for error against radio time, the error recorded and monitored by time announcements.

An inexpensive Time Cube for obtaining *Greenwich Mean Time* or Coordinated Universal Time, as it is now called, can be found in most electronics stores. Some of these receiving devices feature as many as three WWV and WWVH frequencies.

For actual use at sea, I would strongly recommend the acquisition of a suitable quality short-wave receiver. There are many fine receivers on the market today. Find a receiver with as many frequencies for WWV and WWVH as possible. These frequencies are 2.5, 5.0, 10.0, and 15.0 MHz. I am the proud owner of a Zenith Transoceanic Shortwave Receiver. This fine receiver has served me faithfully for many years and I enjoy using it to listen to worldwide broadcasts.

You will require at least one stopwatch. I have two in the event one should fail. I have two of the older hand-held types that you depress to start and again to stop. Any type will do.

For an adequate supply of Line of Position Worksheets go to Appendix D where you will find this form. Taking a few throw-aways into consideration, you will require about twenty-five copies for the course. You could photocopy a generous supply of forms, copied on both sides and hole-punch them for keeping in a binder.

You will need to purchase a star finder. During the course, you will be using the star finder to prepare a list of stars for obtaining a three-star fix, identifying planets, and identifying an unknown star and planet. My star-finder is a Weems & Plath from Annapolis, Maryland identified by its old H.O. No. 2102-D.

After you've gained expertise in utilizing tabular methods of sight reduction and have learned the theory of celestial navigation, you will be ready to use a scientific calculator for sight reduction work and other applications of celestial navigation. Scientific calculators are very inexpensive and range from $10 to $15 or more. There are many manufacturers, such as Casio, Texas Instruments, etc. I have a Radio Shack, LCD scientific model that is satisfactory. It is especially important that the calculator has the three basic trigonometric functions: sin, cos, and tan, and a function button for their respective inverse functions. Other functions like reciprocals, square root, and memory storage are desirable.

Instruments and materials for celestial navigation can be found in chart stores, marine chandlers, and marine mail order catalogs.

MORNING SUN SIGHT -
NORTH PACIFIC OCEAN

TWO

THE MARINE SEXTANT

In order to determine terrestrial positions at sea, it will be necessary to sight the sun, stars, moon, and planets. These observations are made with a marine sextant. So as not to confuse anyone, please remember that a sextant performs only one function: it measures an angle. When a sextant is used for celestial navigation, its primary, but not only, function is to measure an angle called the *altitude* between the celestial body observed and the surface of the sea called the *apparent horizon*.

There is a proper method of taking the sextant out of the box. Reach into the box with your left hand, gently grasping the frame and remove the instrument. The sextant is usually stored with the handle facing downward. With your right hand, grasp the handle so that the telescope is toward your eye. While holding the sextant, compare your instrument with the one illustrated in Figure 2.1. Notice that your sextant is similar and that all the parts in the diagram can be identified on your instrument.

Depress the release lever. See how the *index arm* sweeps from zero degrees (0°) to ninety degrees (90°) or more. Twist the micrometer drum: notice that one full revolution of the drum will move the index arm only one notch or one degree (1°) of arc. Also, notice that the micrometer drum is graduated into sixty parts. Each one of these parts represents one minute (1′), or sixty minutes (60′) equals one degree (1°).

The telescope can be adjusted to obtain a clear image.

The index and horizon shades can be moved into your field of vision by placing them in the line of sight ahead of the mirrors. For now, move the shades out of the field of vision of the mirrors.

The sextant performs its intended function better if it is adjusted to neutralize as many *instrument errors* as possible. Begin by depressing the release lever and move the index arm to where the mark on the index arm corresponds to approximately 50° on the arc.

Sighting from the top of the instrument, as shown in Figure 2.2, see if the *index mirror* is perpendicular to the frame. If the index mirror is not perpendicular to the frame, the arc and the arc's image in the mirror will not appear continuous and unbroken.

You will want to determine whether your sextant is easily adjusted. Adjustments can be made at an instrument shop; however, if your sextant does not require special tools, adjust the instrument yourself. When sailing to many areas of the world, instrument shops may not be available to you, so I recommend learning how to adjust your own sextant. Once you have learned to perform the basic adjustments you will be able to do them in minutes.

My older Ebbco was furnished with a small allen wrench for adjustments. My newer Davis Mark 15 is fitted with three easily adjustable thumbscrews. One screw behind the index mirror and two behind the *horizon glass*. The Davis is by far the easier and faster sextant to adjust since no tools are required.

Some metal sextants I've seen are fitted with protective covers over the thumbscrews. When the covers are removed, the adjusting screws are visible and can be rotated with a tool provided for that

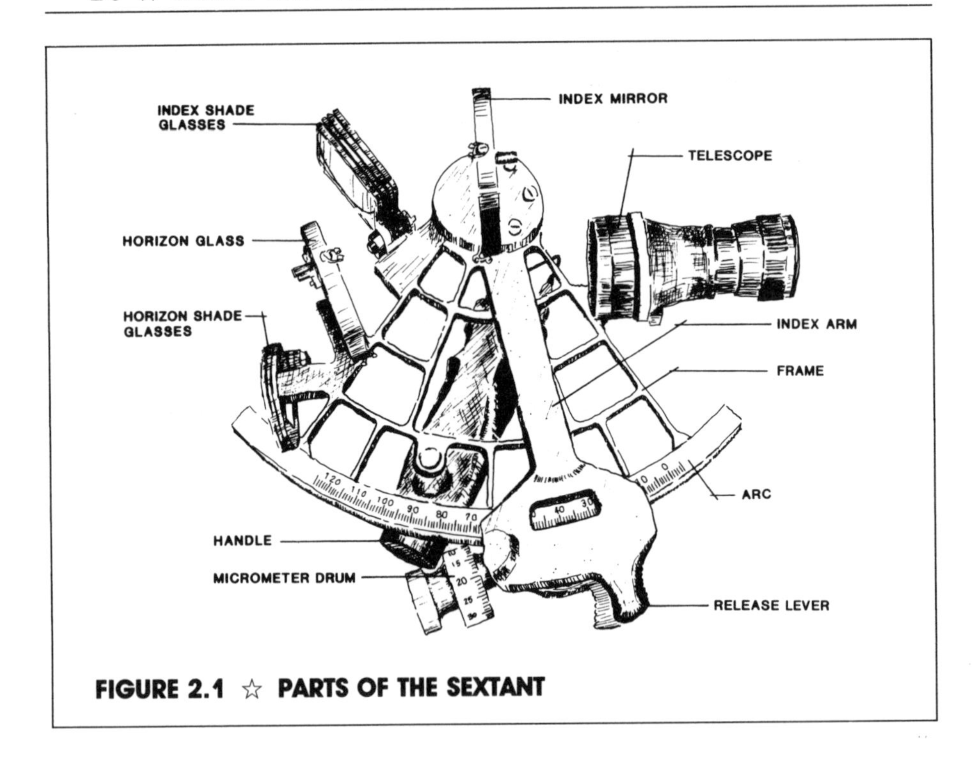

FIGURE 2.1 ☆ PARTS OF THE SEXTANT

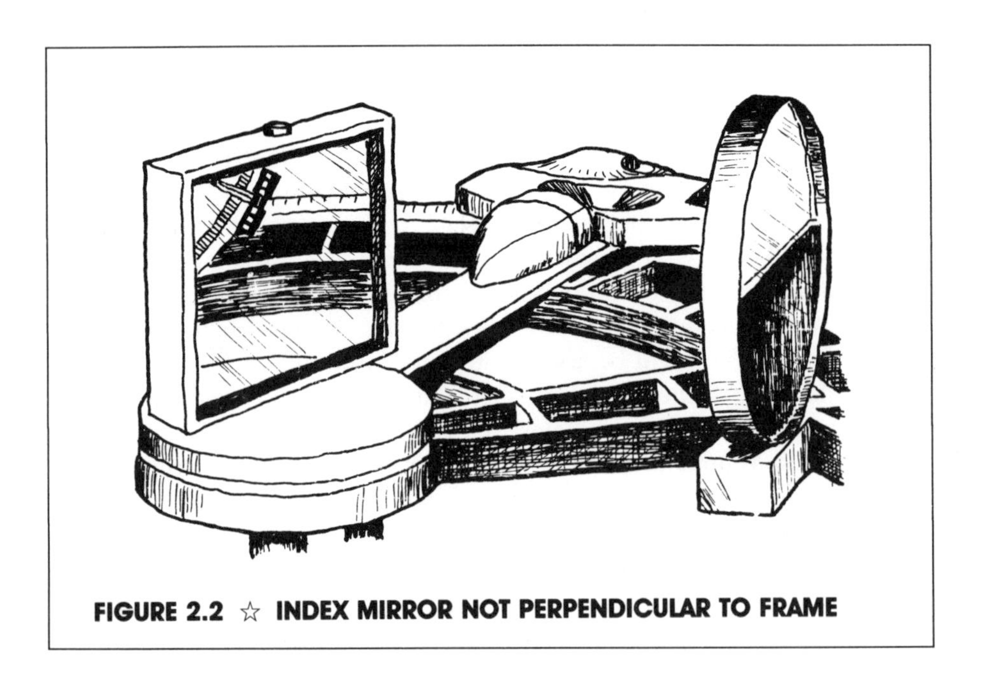

FIGURE 2.2 ☆ INDEX MIRROR NOT PERPENDICULAR TO FRAME

purpose. The tool is usually a small rod that is inserted into holes drilled through the heads of the adjusting screws.

If the arc does not appear in a continuous line as determined above, adjust the index mirror. Rotate the index mirror adjusting screw in either direction until the arc does appear to be continuous. Normally, no further adjustment is required to insure that it is perpendicular to the frame.

The horizon glass now needs to be adjusted so that the glass is perpendicular to the frame and parallel with the index mirror. There are two adjusting screws furnished for this purpose, and they are located behind the horizon glass.

Depress the release lever and set the index arm on approximately 0°.

If your sextant has a lanyard, place the lanyard over your neck (I suggest you add one if it doesn't have one). If you drop the sextant it will fall somewhere above your waist. It's a good habit to always place the lanyard over your neck prior to leaving the cabin with your sextant.

Whether on deck, by the waterfront, or outside your home, you should now peer through the telescope at the apparent horizon or a flat surface in the distance. Slowly rotate the micrometer drum until the apparent horizon or flat surface in the distance appears to be a continuous unbroken line as shown in Figure 2.3.

Without changing the settings on the instrument, find a vertical object in the distance. This might be a mast, flagpole, or the side of a tall building.

Observe the vertical structure through the telescope, and while observing, move the instrument from left to right and from right to left, back and forth. If the vertical object appears duplicated or to skip from one side to the other, the horizon glass is not perpendicular to the frame and requires adjustment. The adjusting screw that is closest to the frame and behind the half-silvered mirror is used to achieve perpendicularity. Rotate the adjusting screw and continue to observe the vertical object while moving the instrument back and forth. When the object no longer appears duplicated or to jump back and forth, the adjustment is completed and the index mirror and the horizon glass will both be perpendicular to the frame. See Figure 2.4.

The most difficult adjustments have been accomplished, and only one remains. Without touching any other adjustments, set the micrometer drum to 0° and again sight through the telescope at the

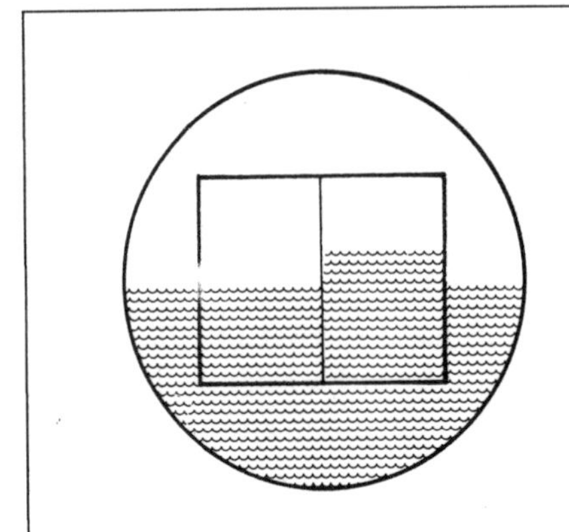

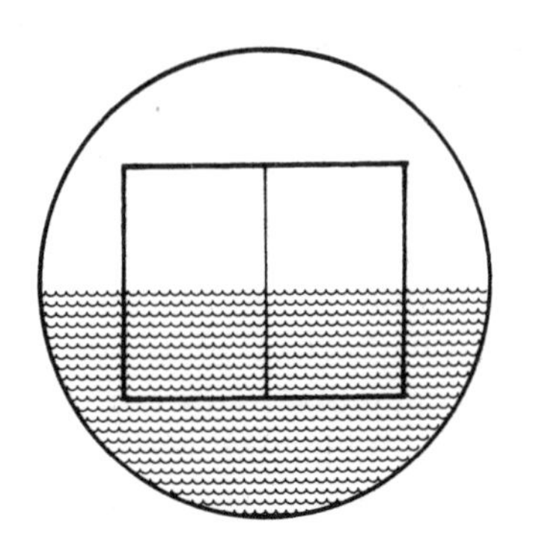

FIGURE 2.3 ☆
HORIZON GLASS NOT ALIGNED WITH HORIZON

HORIZON GLASS IS ALIGNED WITH HORIZON

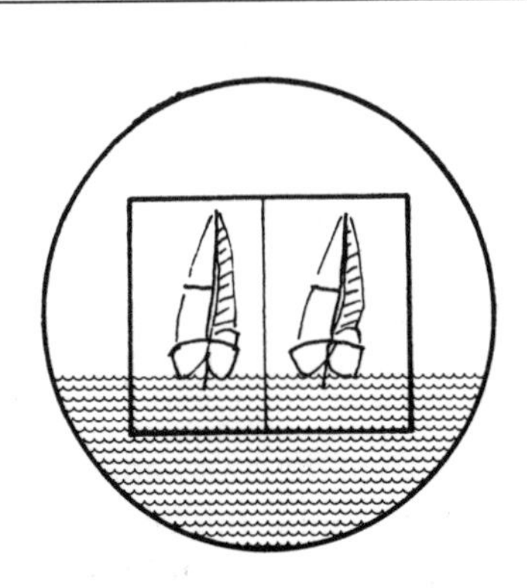

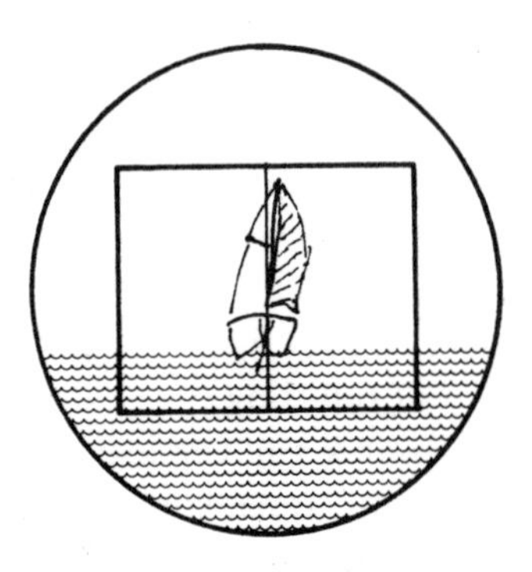

FIGURE 2.4 ☆
HORIZON GLASS NOT PERPENDICULAR TO THE FRAME

HORIZON GLASS IS PERPENDICULAR TO THE FRAME

apparent horizon or a distant flat surface. If there is an *index error* the horizon will not appear to be continuous. The adjusting screw behind the horizon glass and farthest from the frame is for index error adjustment. Rotate the screw until the apparent horizon or flat surface appears continuous. See Figure 2.5.

To be certain that the sextant is in adjustment, sight through the telescope at the horizon or flat surface and slowly rock your head and the sextant in unison. If the horizon or flat surface appears as a continuous, unbroken line, the sextant is in adjustment as shown in Figure 2.6. If the sextant is not in adjustment, repeat all of the above procedures. If it is not possible to adjust the instrument take it to a shop.

If difficulty is encountered in bringing a star or planet to the horizon it is possible that the telescope is not in alignment. Telescope alignment adjustments are done at an instrument shop.

I always adjust my sextants prior to leaving port because, at sea there are no masts, flagpoles, or vertical structures to sight. I do not attempt adjustments while underway and make every effort to keep my instruments from being knocked out of adjustment. I keep them dry, clean the mirrors with fresh water, wipe the mirrors with a lens cloth, and store them in their boxes.

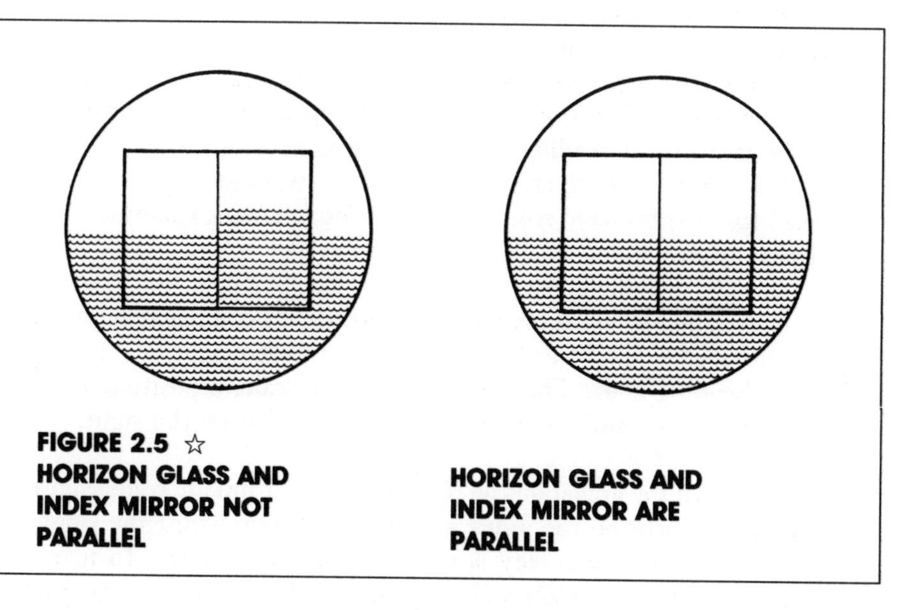

FIGURE 2.5 ☆ HORIZON GLASS AND INDEX MIRROR NOT PARALLEL

HORIZON GLASS AND INDEX MIRROR ARE PARALLEL

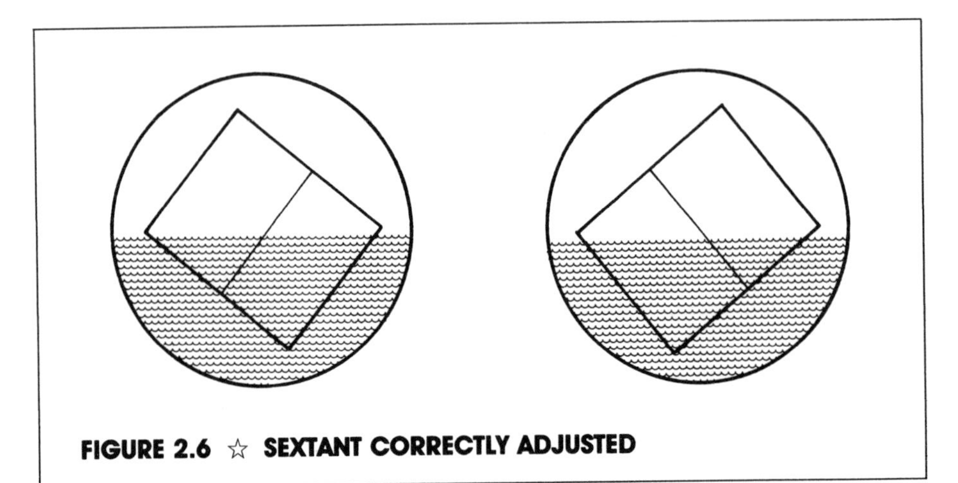

FIGURE 2.6 ☆ SEXTANT CORRECTLY ADJUSTED

If you have a metal sextant, I would suggest the occasional application of light oil to the moving parts.

My sextant is stowed where motion and vibration are minimal.

At sea, some index error normally begins to appear. Index error does not pose a problem other than adding or subtracting the error from the altitude of the observation. However, an index error greater than ±6′ is considered beyond the tolerable limit and a backup sextant should be used.

Index error is determined by viewing the horizon and rotating the micrometer drum until the horizon appears to be continuous. If the instrument reading is 0°00′ there is no index error. If this isn't the case, read the error and determine whether the error is to be added or subtracted from the sight. The following examples will clarify the situation.

1. Reading is 56′. This reading is off the scale, a negative index error of 4′, and the 4′ must be added to the sight.
2. Reading is 4′. This reading is on the scale, a positive index error of 4′, and the 4′ must be subtracted from the sight.

Note that the gradations on the micrometer drum are read against a mark or marks on the index arm itself. If there is just the single mark, instrument accuracy is to one minute (1′) of arc. In many cases, however, a vernier scale is provided that brings accuracy to one or two-tenths (0.1′ or 0.2′) of arc. The micrometer drum is first read to the minute that either lines up exactly with the first mark (the longest) of the vernier scale, or is just outside the vernier scale. The tenths are read on the vernier scale using the mark that exactly lines up with one on the drum.

Read the booklet, and any other information provided with your sextant. The above procedures for adjusting your sextant may have been described in the booklet, and there should be additional information on your particular instrument.

Practice with the sextant by taking sights of the sun. I recommend taking sights from on shore, an anchorage, and the harbor. If you don't live by the water and have an artificial horizon, you can practice taking sights from your backyard. Attempt offshore sights only after you have practiced for awhile.

The following is a procedure for an offshore sight of the sun. The procedure will help you with practice sessions and give you an idea of what you will be attempting to accomplish.

Tune in your shortwave receiver or Time Cube to one of the following frequencies: 2.5, 5.0, 10.0, or 15.0 MHz, and use the frequency that provides optimum reception. These frequencies are for Bureau of Standards Stations WWV and WWVH, and the content of the broadcasts are identical regardless of the frequency selected. In the area of my passages I have found that 15.0 MHz usually provides the best reception during the morning and early afternoon hours. 10.0 MHz usually provides the best reception during the late afternoon and evening hours. In other parts of the world, reception characteristics may vary.

Once a frequency is selected and tuned in, a series of time ticks will be heard. While listening to the time ticks, one or two voices will be heard in the last fifteen seconds of each minute. A woman's voice is broadcast for approximately seven and a half seconds, followed by a man's voice for the remaining seven and a half seconds and terminated by a tone. The announcement is as follows: Female voice: "At the tone, twenty hours, ten minutes, Coordinated Universal Time." Male voice: "At the tone, twenty hours, ten minutes, Coordinated Universal Time." Then the tone is transmitted. At the tone the stopwatch is started and the time announcement: 20h 10m is recorded.

Coordinated Universal Time has replaced the term "Greenwich Mean Time" and it is extremely accurate. The female voice is transmitted from Station WWVH, Kauai, Hawaii and the male voice from Station WWV Fort Collins, Colorado. Which voice or voices you hear depends on your location, time of day, and atmospheric conditions.

Time signals broadcast from stations of other nations may provide clearer reception in distant waters. Find out the relevant frequencies before you leave on a voyage.

Don't forget to set a quartz-crystal wristwatch and/or alarm clock to Greenwich Mean Time. With your radio equipment and several inexpensive clocks you are provided with a more than adequate backup.

Before you take a sight, determine a good location on deck from which it can be made. I recommend that you be as high above water as possible, and you should be able to attach your safety harness to a secure point with a lanyard.

Now for the sun sight itself. Hand the stopwatch to your mate, or if single-handing, put it in a pocket or hang it on your neck with a lanyard.

Take the sextant from its box, and move one of the darker-tinted shades into the line of vision between the horizon glass and the index mirror. Get yourself set in your pre-selected deck location. Estimate the angle between the sun and the horizon, and set the index arm by depressing the release lever and moving the arm to the desired degree mark. View the horizon in a sweeping manner until the sun is observed in the telescope. Now depress the release lever and move the index arm until the sun is very nearly on the horizon.

Be certain that you have provided adequate shading to protect your eyes from the harmful rays of the sun, and that the sun appears as a well-defined ball in the telescope. Experiment with the shading until you are comfortable and satisfied with it.

With the sun near the horizon, let go the release lever to clamp the index arm, and transfer your fingers to the micrometer drum. Rotate the drum while rocking your head and the sextant in unison, until the sun's lower limb (the very bottom of the sun) barely touches the horizon at the bottom of the pendulum-type arc. This technique is called *swinging the arc*. When the sun touches the horizon at the bottom of the arc, the command "Mark" is given, and the stopwatch is stopped.

While carrying out the technique described above, there are several things to consider:

1. The index arm is moved to bring the sun as near to the horizon as possible. If there is too much glare on the horizon, a shade can be moved into the field of vision.

2. The sun is viewed through the telescope, head and instrument rocked in unison, micrometer drum rotated, and the command "Mark" is given at the instant the lower limb of the sun barely touches the horizon at the bottom of the inscribed arc, as shown in Fig. 2.7.
3. The horizon must appear as a straight, continuous, horizontal line. If the horizon does not appear as a straight, continuous, horizontal line it is because the craft is sailing in a trough and a wave is obstructing the horizon. Observation is continued until the craft is on top of a wave.
4. The settings on the sextant are not moved and the stopwatch is not reset until the readings are recorded.
5. Do not get discouraged as it doesn't take long to learn how to obtain acceptable observations.
6. Never use the sextant to fend off a shroud or for anything but its intended purpose. If both hands are needed for any reason, let the sextant fall harmlessly to the waist. This is why a lanyard on the sextant and around your neck is so important.

Returning to the command: "Mark." This was the precise instant when the lower limb of the sun barely touched the horizon. If your mate had not been available to stop the stopwatch, or if you were single-handing, you might have counted to five while digging the stopwatch out of your pocket and stopping it. The single-handed procedure works best if you remember to subtract the five seconds from the stopwatch reading when it is recorded.

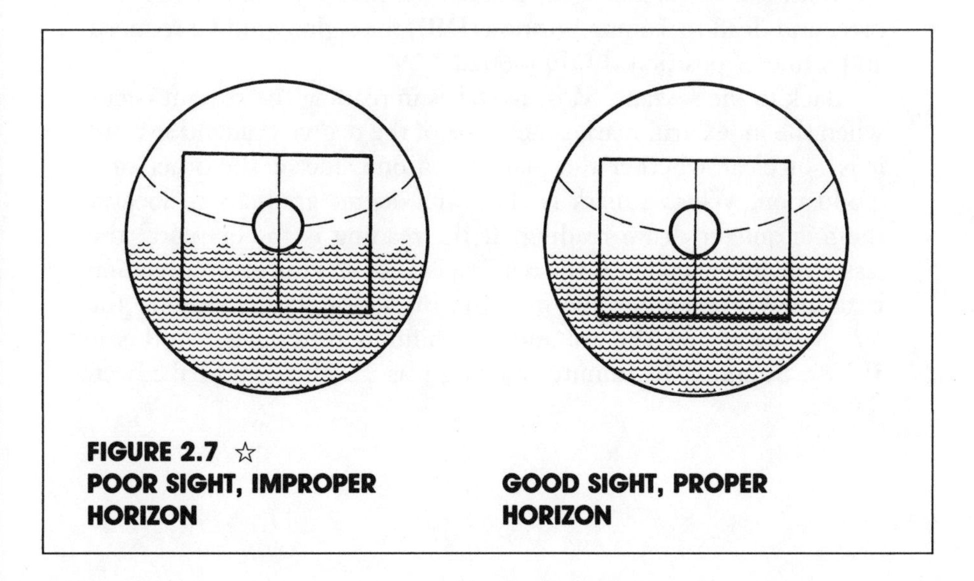

FIGURE 2.7 ☆
POOR SIGHT, IMPROPER HORIZON **GOOD SIGHT, PROPER HORIZON**

Suppose the stopwatch time was 3m 25s, and 5s was subtracted for the delay. The exact time of the sight described above would have been:

WWV	20h 10m
Stopwatch	3m 20s
GMT	20h 13m 20s

Assume that prior to this sight index error was checked. The reading obtained from the mark on the index arm corresponding to the micrometer drum was 58′. This means there was a *negative* index error of 2′ and it had to be added to the sight.

The mark on the index arm appears to be between the 50° and 60° graduations on the arc. Counting from the 50° side toward 60° there are three lines and the mark appears to be somewhere between the third and fourth lines. 53° is recorded.

Next, look at the index mark corresponding to the micrometer drum. The number closest to the longer mark on the index arm is somewhere between 25′ and 30′. Counting from 25′ toward 30′ between two and three marks . . . closer to three marks so the reading is between 27′ and 28′, closer to 28′, and is rounded off and recorded as 28′. Micrometer drum readings to the nearest minute are close enough for small-craft navigation.

The *sextant altitude* (Hs) obtained is 53°28′ and the index correction (IC) of 2′ is added to Hs. Other corrections are applied to further correct the sight and will be discussed in the next chapter.

With the corrected sight, Greenwich Mean Time (GMT), the date, and dead reckoning position (DR), the sight could be reduced and a line of position (LOP) plotted.

Back to the sextant. Most mistakes in reading the sextant occur when the index arm mark is near one of the degree graduations and it is not clear whether the mark is on one side or the other of a graduation. When a mark is close to a degree graduation, look at the micrometer drum reading. If the reading is 55′ or more, the lesser reading is taken. If the reading is 5′ or less the greater reading is taken. For example: the mark on the index arm is close to the 60° graduation on the arc and the minutes reading is 58′. Hs is 59°58′. Suppose the minutes reading was 2′. Hs would have been 60°02′.

DEPARTING - MORRO BAY, CALIF.

THREE

THE SUN SIGHT

Let's get into the fun part as you sail with us from Morro Bay, California to the Hawaiian Islands. The date is August 31, 1986. My wife Enid and I have just had a great breakfast and are enjoying a going-away party with our relatives and close friends. At approximately 1045 we vacate our slip in Morro Bay, wave goodbye to everyone, and at about 1100 we pass through the harbor entrance. The harbor entrance is adjacent to Morro Rock, which is often referred to as the Gibralter of the Pacific. After many months of preparation, we are finally outward bound for the Hawaiian Islands. The wind is light and we steer a course of 252° true as we motorsail into the fog. Within minutes Morro Rock and all other landmarks disappear as we are engulfed by the fog.

I begin the plot of our passage.

Early in the evening the wind begins to pipe up. I am grateful that my past experience has taught me to begin voyages from the California coast with reduced sails. We are carrying the smaller working jib and a reef in the mainsail. The wind commences to blow out of the northwest at about twenty-five knots, and maintains its velocity for most of the evening as we claw our way offshore on a close reach. Toward morning we are motorsailing, and the coastal fog has disappeared. We are able to see the sun intermittently between cloud cover and we decide to determine our position.

Begin by removing a *Universal Plotting Sheet* from the pad you purchased. From Morro Bay you will be plotting the course to our present dead reckoning (DR) position. This position is based on the course steered, the speed through the water, and the elapsed time. On sailboat passages, the distance (which is the product of time and speed) is measured by a log, and this, rather than speed through the water, is shown on the course plot. Our course from Morro Bay (Lat. 35°22′ N, Lo. 120°52′ W) is 252° True, and the ZOE's log reading is 95 nautical miles at 1100, when we take the first sight of the sun. In other words, we sailed the 95-mile distance in the last 24 hours.

Begin by observing the layout of the plotting sheet. Note that there are five equally-spaced horizontal lines. These represent *latitude*. On the right side, from top to bottom, label the latitudes 36°, 35°, 34°, 33°, and 32° as shown in Fig. 3.1. Note that each degree of latitude is shown as 60′ on the meridian running from the bottom to the top of the sheet in the center. Each minute on the scale in the center of the sheet is one nautical mile, and all distances are measured from this scale. Note that the scale printed in the lower right hand corner is labeled MID-LATITUDE and LONGITUDE SCALE. This is used for scaling the *longitude* with respect to the closest latitude, and for plotting the *meridians* of longitude.

The Mid-Lat. and Long. scale is never used for measuring distances. The only location on the earth's surface where longitude is

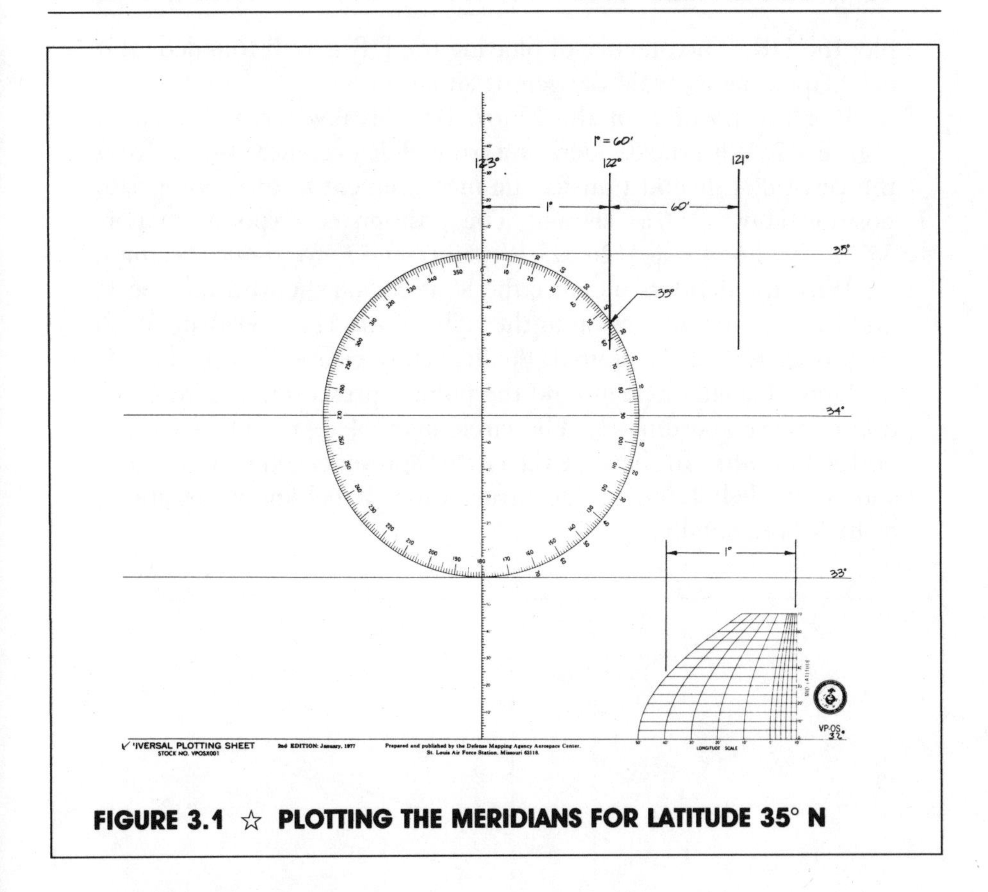

FIGURE 3.1 ☆ PLOTTING THE MERIDIANS FOR LATITUDE 35° N

equal to latitude is on the equator. North and south of the equator the meridians become progressively closer together, and join at their respective poles.

You will want to plot the meridians at Lat. 35° N, corresponding to Long. 121°, 122°, and 123° W. At 35° N, place one pin of the dividers on the left side of the scale and the other pin horizontally on the extreme right side of the scale. Transfer the dimension obtained from the dividers to the center of the plotting sheet, on the 35° latitude line twice, as shown. Using your parallel rulers, or other suitable tool, draw in the two meridians and label them from right to left 121°, 122°, and 123°. Thus, a convenient portion of the chart has been created in a scale that is easy to work with.

Another technique would be to draw the meridian closest to the center of the plotting sheet through the compass rose at 35° as shown. The other meridians could then be drawn at a distance equal to the first.

The DR is plotted and labelled next. It is always important to plot the DR. The practice of plotting the DR is well-founded, and has helped me to avoid dangerous situations.

Plot the position of the Morro Bay breakwater as shown in Figure 3.2. With the dividers or a parallel device, measure 22′ from the vertical scale and transfer the measurement to the appropriate position above 35° as shown. The position corresponds to Lat. 35°22′ N. For Long. 120°52′ W, subtract 52′ from 60′ obtaining 8′. With the dividers measure the 8′ at 35° on the Mid-Lat. Scale and transfer the dimension to the right of the 121° meridian. With the Long. 120°52′ W plotted, the departure position is established.

Draw a small circle around the point representing the intersection of these coordinates. The circle symbol represents a known position or a fix. In this case the circle represents a known position and is labelled: 1100, our departure time. Label known positions or fixes horizontally.

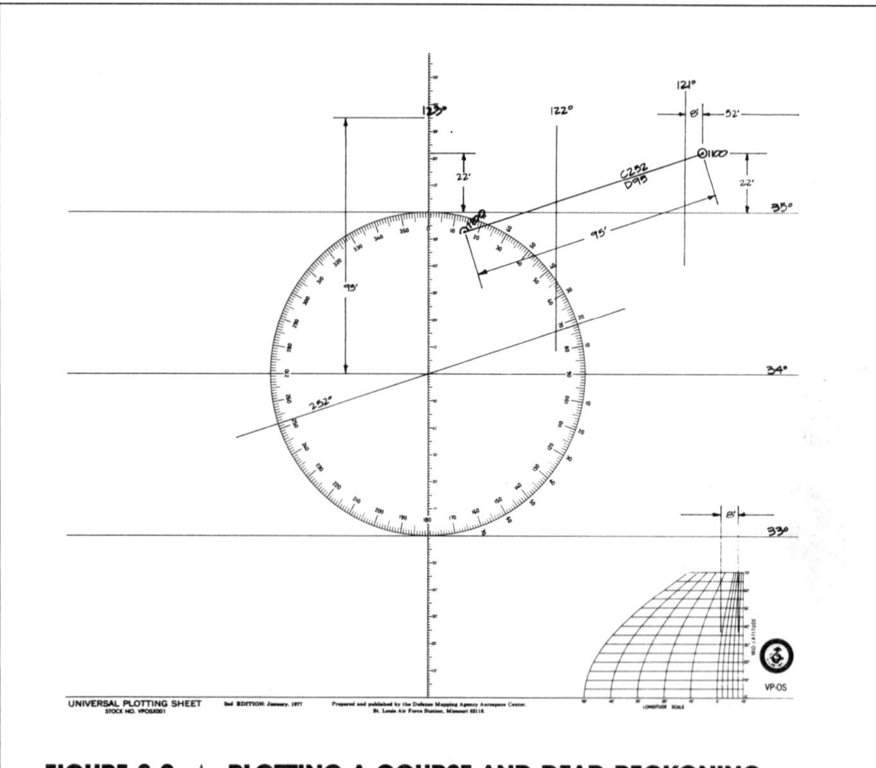

FIGURE 3.2 ☆ PLOTTING A COURSE AND DEAD RECKONING POSITION (DR)

Transfer the course of 252° from the true compass in the center of the plotting sheet with parallel rulers, parallel plotter, or a protractor.

Draw the DR course line and measure 95 nm along it from the vertical scale. Conveniently, one inch corresponds to twenty nautical miles or 20′ on the vertical scale.

Label the course above the line: C252 and the distance: D95 below the line. When plotting DRs for celestial navigation the true, 360° course is shown. Draw a half circle around the point representing the 95 nautical miles and label it: 1100, the approximate time of the first sight. The label is drawn at an angle of approximately 45°, the proper convention for labelling dead reckoning positions.

For larger ships and power boats the speed is usually shown below the course line. For example: S10 would indicate a speed of ten knots. Since the speed of a sailing yacht varies considerably, dependent upon wind and sea conditions, the convention of showing the total distance covered is proper and considerably more useful.

Determine the latitude and longitude of the DR position you have plotted. With your dividers, take the the DR Lat. from the vertical scale, and the DR Long. from the Mid-Lat. scale. I figured the latitude and longitude of the DR position to be: Lat. 34°53′ N, Long. 122°43′ W.

The DR is where we think we might be after sailing and powering for 24 hours. The DR is used as a starting point to determine the actual terrestrial position by celestial navigation.

Begin by entering the following information on a LINE OF POSITION WORKSHEET. In the spaces provided, record: No. 1; Area: CENTRAL CALIFORNIA COAST; Body: SUN; Date: 9/1/86; D.R. Lat.: 34°53′ N; D.R. Long.: 122°43′ W; Log: 95; and Local Time: 1100 as shown in Figure 3.3.

My wristwatch was set on Greenwich Mean Time (GMT). I started my wristband stopwatch at 18h 59m GMT. Record: WWV—18h 59m in the space near the top of the form.

I removed the sextant from the box, placed the lanyard over my neck, and armed with sextant and started stopwatch I emerged reluctantly and a little queasily (first day out) from the cabin to begin my daily business of navigating.

On deck, my first duty and obligation to my well being was to firmly attach myself to the boat via my safety harness. I determined

NO. 1 — **LINE OF POSITION WORKSHEET** — AREA CENTRAL CALIFORNIA COAST

BODY	DATE	D. R. LAT.	D. R. LONG.	LOG	WWV	STOP WATCH	LOCAL TIME
SUN	9-1-86	N 34° 53'	W 122° 43'	95	18h 59m	1m 12s	1100

TIME	HR.	MIN.	SEC.
WWV	18	59	—
Stop Watch	—	1	12
GMT	19	0	12

ASS. POSITION LHA (ALMANAC)

GHA Hours		105° 00.7'
Min. & Sec.		0° 03.0'
Stars Only SHA		—
Moon & Planets Only V:	Corr.	—
360°		360°
Total GHA		465° 04'
Ass. Long.		123° 04'
LHA		342°

DECLINATION (ALMANAC)

Dec. Hours		N 8° 11.0'
"d" ± 0.9	Corr. ±	0.0'
True Dec.		N 8° 11'

SUN, STARS, & PLANETS SEXTANT (ALMANAC)

Hs	58° 48'	Ha	58° 44.9'
IC	—	Corr.	15.4'
Dip	−3.1'	Add'l Corr.	—
Ha	58° 44.9'	Ho	59° 00'

INTERCEPT AZIMUTH – FIRST ENTRY

Lat. 35	LHA 342	A 14° 40'	A° 15	A' 40
		B ⊕ 53° 38'	Z1 ⊕ 79.4°	
		Dec. ⊕ 8° 11'		
	Sum: B+Dec.	F 61° 49'	F° 62	F' 49

INTERCEPT AZIMUTH – SECOND ENTRY

A° 15	F° 62	H 58° 31'	P° 60	Z2 ⊕ 64.0°
F' 49	P° 60	Corr.1 ± 10'	Z2° 64	
	Sum	58° 21'	Z1 79.4°	
A' 40	Z2° 64	Corr.2 ⊕ 09'	Z2 64.0°	
	True Hc	58° 30'	Z 143°	
	True Ho	59° 00'	Zn 143°	
	Intercept	30'	T ✓	A

MOON ONLY SEXTANT (ALMANAC)

Corr.		Hs
Add'l Corr.		IC
HP	Corr.	Dip
Total		Ha
	Total	
	Ho	

PLOTTING

Ass. Long.	W 123° 04'
Zn 143°	Ass. Lat. N 35°
Intercept	T 30' A

FIGURE 3.3 ☆

that my best position for this sight would be from the stern rail port side. While the sea was not rough, there was a long and fairly high swell condition.

From my position I observed the sun slipping in and out of the clouds. I found a couple of index shades that made the sun appear as a well defined ball. I checked the sextant for index error. The sextant had been adjusted to zero index error prior to departure and no index error had started to appear. I released the index arm and preset the sextant at about 60°. Looking through the telescope I began to search the horizon and found the sun at approximately 59°. There was no need to shade the horizon glass as the horizon was clear and sharp. I continued to steady myself, rocking the sextant and rotating the micrometer drum until the lower limb of the sun just touched the horizon at the bottom of its arc. I yelled "mark" to myself (my wife felt even queasier and was unwilling to participate), counted to five seconds and stopped the stopwatch on my wristband.

I went below to reduce the sight. I was especially careful not to touch the stopwatch or sextant settings. The stopwatch read: 1m 17s. Subtract the 5s it took to stop the stopwatch, record: 1m 12s in the space near the top of the form.

Record the sextant reading in the block Sun, Stars, and Planets Sextant (Almanac) in the space labelled: Hs. The index arm degrees mark was between 58° and 59°, closer to 59° on the arc, and the index arm minutes mark was very nearly on 48′ on the micrometer drum. Record: Hs: 58°48′ in the space provided.

There was no *personal error*. Personal error is incurred by observations of a particular navigator. An example of personal error would be the tendency of an observer to oversight the horizon and always measure Hs 2′ greater than other observers. The 2′ error would be subtracted before recording Hs.

To determine if you have a personal error, take as many sights as possible from a known location, and work them out. If the results are consistently a little over or a little under the actual figure, this is your personal error. Since in this case there was no index error or personal error, put a slash through the next space, labelled IC.

Beginning with the Time block record: WWV; 18h 59m—s; Stop Watch:—h 1m 12s. Add these times to obtain: GMT: 19h 0m 12s. Note that in the *Nautical Almanac* extracts used in this book, time is referred to as GMT with UT in parentheses. Starting in 1989, the *Nautical Almanac* will refer to time as UT, with GMT in parentheses. In future years the GMT reference will be dropped.

Correct the sextant altitude (Hs) for *dip*, *refraction*, *atmospheric pressure*, and *temperature*. Don't let all these terms bother you as the corrections are easily accomplished by using information contained in the inside front cover and first two pages of the *Nautical Almanac*. Some of this information is duplicated on a white card that is provided with the almanac. The inside front cover and white card are headed: A2 ALTITUDE CORRECTION TABLES 10°–90° —SUN, STARS, PLANETS. Use your current year almanac. This data is the same every year except for some additional corrections under the heading: STARS AND PLANETS, for the planets only.

Go to the column headed: DIP on the right side of the inside cover or white card. The dip correction is to compensate for the height of an observer's eye above the water. My eye was ten feet above the water. Go down the column: Ht. of Eye. in ft. between 9.8 and 10.5 feet. Go to the left and find the correction under the column headed: Corr. and record: Dip: −3.1′. The dip correction is always negative, and is subtracted from the sextant altitude. Subtracting the dip from the sextant altitude we obtain and record: Ha: 58°44.9′ and rewrite Ha, the *apparent altitude* (Ha), in the upper right hand space.

Go to the first two columns headed: OCT.–MAR. SUN APR.–SEPT. Go across to the second column headed: APR.–SEPT., go down the App. Alt. column and stop between 57°02′ and 61°51′ corresponding to Ha: 58°44.9′. Take the correction from the column headed: Lower Limb. The *lower limb* correction is always positive. On the form the correction is recorded in the space: Corr.; 15.4′. If we had taken a sight of the *upper limb* of the Sun we would have used the column headed: Upper Limb. The upper limb correction is always negative. I have rarely taken a sight of the upper limb of the Sun. The upper limb is more frequently used for low altitude sights of less than 10°.

There are two more pages of corrections. The next page is headed: ALTITUDE CORRECTION TABLES 0°–10°—SUN, STARS, PLANETS A3. These corrections are a continuation of the sun, stars, and planets corrections from the inside front cover for low altitude sights of 10° or less. They are not needed for this sight reduction.

The following page is headed: A4 ALTITUDE CORRECTION TABLES—ADDITIONAL CORRECTIONS. ADDITIONAL REFRACTION CORRECTIONS FOR NON-STANDARD CONDITIONS. The non-standard corrections are for high or low temperatures versus changes in atmospheric pressure. These corrections have no effect on high altitude sights above 50°, and the greatest effect on low altitude sights below 10°. Put a slash in the space: Add'l Corr.

Add Ha to Corr. obtaining the *observed altitude* (Ho). Round off Ho to the nearest minute and record: Ho: 59°00′.

You are ready to use the contents of the *Nautical Almanac*. The almanac is published annually. The almanac is organized a little like some telephone books with white and yellow pages. The similarity, however, ends there.

The main portion of the white pages contains the records of celestial positions for each hour of the year, and information regarding twilight times, etc. The remaining few pages of the white pages contain the sight reduction tables we will become intimately familiar with. With exception of the sight reduction tables, most of the information in the white pages changes each year.

The yellow pages contain the additional angular changes required for the remaining minutes and seconds, conversions of arc to time, some interpolating values, and with exception of the section headed: INDEX TO SELECTED STARS—1986, they do not change from year to year.

Go to the right hand white page of the 1986 almanac headed: 1986 SEPTEMBER 1, 2, 3 (MON., TUES., WED.) of Figure 3.4. This is a typical right hand page of the almanac.

The first column headed G.M.T./(UT) corresponds to each hour of GMT for each of the 24 hours of the three days shown. From the time block on the form you have 19h GMT. Go down the first column, d, 1, MONDAY, to 19h. Go across to the next column headed: SUN. There are two separate columns of information for the Sun.

The first column is headed: G.H.A. for *Greenwich hour angle.* Take G.H.A. from this column and on the worksheet record it in the space: GHA Hours: 105°00.7′.

The second column is headed: Dec. for *declination.* Take Dec. from this column and record it on the worksheet in the space: Dec. Hours: 8°11.0′ N.

Go to the bottom of the Dec. column for the three days and find a number called little d. In the space under the Dec. Hours space record: "d" ±: 0.9 and ask yourself this question. Is the declination increasing or decreasing? During the next day the declination decreased from N 8° to N 7° and the minutes are also decreasing. Since the declination is decreasing, circle the − sign. The − sign is to be applied to the correction for the change in the minutes of declination and the correction subtracted from the Dec. Hours. If the declination were increasing, the + sign would have been circled and the correction added to the Dec. Hours.

Go to the yellow pages headed: INCREMENTS AND CORRECTIONS to obtain the minutes and seconds corrections to GHA and the little d correction for Dec. Hours.

With the GMT Min. & Sec. taken from the Time Block go to the Om and lm page. Go down the first half of the page and down the extreme left column headed: Om and stop at 12s. Go across to the first column headed: SUN/PLANETS and record in the space on the worksheet headed: Min. & Sec.: 0°03.0′.

On the same half page for Om there are three columns for v or d corrections. Go down the v or d columns and stop at 0.9 corresponding to little d. Next to the v or d columns are corresponding

1986 SEPTEMBER 1, 2, 3 (MON., TUES., WED.) 173

G.M.T. (UT)	SUN		MOON				
d h	G.H.A. ° ′	Dec. ° ′	G.H.A. ° ′	*v* ′	Dec. ° ′	*d* ′	H.P. ′
1 00	179 56.9	N 8 28.2	217 07.2	10.4	N25 15.3	07.1	55.1
01	194 57.1	27.3	231 36.6	10.4	25 08.2	07.1	55.1
02	209 57.3	26.4	246 06.0	10.5	25 01.1	07.4	55.1
03	224 57.5	. . 25.5	260 35.5	10.5	24 53.7	07.4	55.2
04	239 57.7	24.6	275 05.0	10.6	24 46.3	07.6	55.2
05	254 57.9	23.7	289 34.6	10.6	24 38.7	07.7	55.2
06	269 58.1	N 8 22.8	304 04.2	10.6	N24 31.0	07.8	55.2
07	284 58.3	21.9	318 33.8	10.7	24 23.2	07.9	55.2
08	299 58.5	21.0	333 03.5	10.7	24 15.3	08.1	55.3
M 09	314 58.7	. . 20.0	347 33.2	10.8	24 07.2	08.2	55.3
O 10	329 58.9	19.1	2 03.0	10.8	23 59.0	08.3	55.3
N 11	344 59.1	18.2	16 32.8	10.9	23 50.7	08.4	55.3
D 12	359 59.3	N 8 17.3	31 02.7	10.8	N23 42.3	08.6	55.3
A 13	14 59.5	16.4	45 32.5	11.0	23 33.7	08.7	55.4
Y 14	29 59.7	15.5	60 02.5	11.0	23 25.0	08.7	55.4
15	44 59.9	. . 14.6	74 32.5	11.0	23 16.3	09.0	55.4
16	60 00.1	13.7	89 02.5	11.1	23 07.3	09.0	55.4
17	75 00.3	12.8	103 32.6	11.1	22 58.3	09.1	55.5
18	90 00.5	N 8 11.9	118 02.7	11.1	N22 49.2	09.3	55.5
19	105 00.7	11.0	132 32.8	11.2	22 39.9	09.4	55.5
20	120 00.9	10.1	147 03.0	11.3	22 30.5	09.5	55.5
21	135 01.1	. . 09.2	161 33.3	11.3	22 21.0	09.6	55.5
22	150 01.3	08.2	176 03.6	11.3	22 11.4	09.7	55.6
23	165 01.5	07.3	190 33.9	11.4	22 01.7	09.9	55.6
2 00	180 01.6	N 8 06.4	205 04.3	11.4	N21 51.8	09.9	55.6
01	195 01.8	05.5	219 34.7	11.5	21 41.9	10.1	55.6
02	210 02.0	04.6	234 05.2	11.5	21 31.8	10.2	55.7
03	225 02.2	. . 03.7	248 35.7	11.6	21 21.6	10.3	55.7
04	240 02.4	02.8	263 06.3	11.6	21 11.3	10.4	55.7
05	255 02.6	01.9	277 36.9	11.7	21 00.9	10.5	55.7
06	270 02.8	N 8 01.0	292 07.6	11.7	N20 50.4	10.6	55.7
07	285 03.0	8 00.1	306 38.3	11.7	20 39.8	10.7	55.8
T 08	300 03.2	7 59.1	321 09.0	11.8	20 29.1	10.8	55.8
U 09	315 03.4	. . 58.2	335 39.8	11.9	20 18.3	10.9	55.8
E 10	330 03.6	57.3	350 10.7	11.9	20 07.4	11.1	55.8
S 11	345 03.8	56.4	4 41.6	11.9	19 56.3	11.1	55.9
D 12	0 04.0	N 7 55.5	19 12.5	12.0	N19 45.2	11.2	55.9
A 13	15 04.2	54.6	33 43.5	12.0	19 34.0	11.4	55.9
Y 14	30 04.4	53.7	48 14.5	12.1	19 22.6	11.4	55.9
15	45 04.6	. . 52.8	62 45.6	12.1	19 11.2	11.5	56.0
16	60 04.8	51.9	77 16.7	12.2	18 59.7	11.7	56.0
17	75 05.0	50.9	91 47.9	12.2	18 48.0	11.7	56.0
18	90 05.2	N 7 50.0	106 19.1	12.2	N18 36.3	11.9	56.0
19	105 05.4	49.1	120 50.3	12.3	18 24.4	11.9	56.1
20	120 05.6	48.2	135 21.6	12.3	18 12.5	12.0	56.1
21	135 05.8	. . 47.3	149 52.9	12.4	18 00.5	12.1	56.1
22	150 06.0	46.4	164 24.3	12.4	17 48.4	12.2	56.1
23	165 06.2	45.5	178 55.7	12.5	17 36.2	12.3	56.2
3 00	180 06.4	N 7 44.5	193 27.2	12.5	N17 23.9	12.4	56.2
01	195 06.6	43.6	207 58.7	12.5	17 11.5	12.5	56.2
02	210 06.8	42.7	222 30.2	12.6	16 59.0	12.6	56.2
03	225 07.0	. . 41.8	237 01.8	12.7	16 46.4	12.6	56.2
04	240 07.2	40.9	251 33.5	12.6	16 33.8	12.8	56.3
05	255 07.5	40.0	266 05.1	12.7	16 21.0	12.8	56.3
06	270 07.7	N 7 39.0	280 36.8	12.8	N16 08.2	12.9	56.3
W 07	285 07.9	38.1	295 08.6	12.8	15 55.3	13.0	56.3
E 08	300 08.1	37.2	309 40.4	12.8	15 42.3	13.1	56.4
D 09	315 08.3	. . 36.3	324 12.2	12.8	15 29.2	13.2	56.4
N 10	330 08.5	35.4	338 44.0	12.9	15 16.0	13.2	56.4
E 11	345 08.7	34.5	353 15.9	13.0	15 02.8	13.4	56.4
S 12	0 08.9	N 7 33.5	7 47.9	12.9	N14 49.4	13.4	56.5
D 13	15 09.1	32.6	22 19.8	13.0	14 36.0	13.5	56.5
A 14	30 09.3	31.7	36 51.8	13.1	14 22.5	13.5	56.5
Y 15	45 09.5	. . 30.8	51 23.9	13.0	14 09.0	13.7	56.5
16	60 09.7	29.9	65 55.9	13.1	13 55.3	13 7	56.6
17	75 09.9	29.0	80 28.0	13.2	13 41.6	13.8	56.6
18	90 10.1	N 7 28.0	95 00.2	13.1	N13 27.8	13.8	56.6
19	105 10.3	27.1	109 32.3	13.2	13 14.0	14.0	56.6
20	120 10.5	26.2	124 04.5	13.3	13 00.0	14.0	56.7
21	135 10.7	. . 25.3	138 36.8	13.2	12 46.0	14.1	56.7
22	150 10.9	24.4	153 09.0	13.3	12 31.9	14.1	56.7
23	165 11.1	23.4	167 41.3	13.3	12 17.8	14.2	56.7
	S.D. 15.9	*d* 0.9	S.D.	15.1		15.2	15.4

Lat.	Twilight Naut.	Twilight Civil	Sunrise	Moonrise 1	Moonrise 2	Moonrise 3	Moonrise 4
°	h m	h m	h m	h m	h m	h m	h m
N 72	////	02 36	04 04	▭	▭	00 16	03 29
N 70	////	03 03	04 18	▭	▭	01 20	03 49
68	01 37	03 23	04 28	▭	▭	01 55	04 04
66	02 13	03 39	04 37	▭	00 06	02 20	04 16
64	02 39	03 52	04 44	24 44	00 44	02 39	04 27
62	02 58	04 02	04 51	25 10	01 10	02 55	04 35
60	03 13	04 11	04 56	25 31	01 31	03 08	04 43
N 58	03 26	04 19	05 01	00 18	01 47	03 19	04 49
56	03 37	04 26	05 05	00 37	02 02	03 28	04 55
54	03 46	04 32	05 09	00 53	02 14	03 37	05 00
52	03 54	04 37	05 13	01 07	02 24	03 44	05 05
50	04 01	04 42	05 16	01 19	02 34	03 51	05 09
45	04 16	04 52	05 23	01 44	02 54	04 06	05 18
N 40	04 27	05 00	05 28	02 04	03 10	04 17	05 25
35	04 37	05 07	05 33	02 21	03 24	04 27	05 31
30	04 44	05 13	05 37	02 35	03 35	04 36	05 37
20	04 56	05 22	05 44	02 59	03 55	04 51	05 47
N 10	05 05	05 29	05 51	03 20	04 13	05 04	05 55
0	05 11	05 36	05 56	03 39	04 29	05 17	06 03
S 10	05 16	05 41	06 02	03 59	04 45	05 29	06 11
20	05 20	05 46	06 08	04 19	05 02	05 42	06 19
30	05 23	05 51	06 15	04 43	05 22	05 56	06 28
35	05 24	05 53	06 18	04 57	05 33	06 05	06 34
40	05 24	05 55	06 23	05 13	05 46	06 15	06 40
45	05 24	05 58	06 27	05 32	06 02	06 26	06 47
S 50	05 23	06 01	06 33	05 56	06 21	06 39	06 55
52	05 23	06 02	06 36	06 08	06 29	06 46	06 59
54	05 22	06 03	06 39	06 21	06 39	06 53	07 03
56	05 22	06 04	06 42	06 35	06 50	07 00	07 08
58	05 21	06 06	06 45	06 53	07 03	07 09	07 13
S 60	05 19	06 07	06 49	07 14	07 18	07 19	07 19

Lat.	Sunset	Twilight Civil	Twilight Naut.	Moonset 1	Moonset 2	Moonset 3	Moonset 4
°	h m	h m	h m	h m	h m	h m	h m
N 72	19 52	21 18	////	▭	22 15	20 41	19 56
N 70	19 39	20 52	23 23	▭	21 10	20 19	19 47
68	19 29	20 33	22 15	▭	20 33	20 01	19 39
66	19 20	20 18	21 41	20 41	20 07	19 47	19 32
64	19 13	20 05	21 17	20 02	19 47	19 36	19 26
62	19 07	19 55	20 59	19 35	19 30	19 26	19 21
60	19 02	19 46	20 44	19 14	19 16	19 17	19 17
N 58	18 57	19 39	20 31	18 56	19 04	19 09	19 13
56	18 53	19 32	20 21	18 42	18 54	19 03	19 09
54	18 49	19 26	20 12	18 29	18 45	18 56	19 06
52	18 46	19 21	20 04	18 18	18 36	18 51	19 03
50	18 43	19 16	19 57	18 08	18 29	18 46	19 01
45	18 36	19 06	19 42	17 46	18 13	18 35	18 55
N 40	18 31	18 58	19 31	17 29	18 00	18 26	18 50
35	18 26	18 52	19 22	17 15	17 48	18 18	18 46
30	18 22	18 46	19 15	17 02	17 38	18 11	18 42
20	18 15	18 37	19 03	16 40	17 21	17 59	18 36
N 10	18 09	18 30	18 55	16 21	17 06	17 49	18 30
0	18 03	18 24	18 48	16 03	16 52	17 39	18 24
S 10	17 58	18 19	18 43	15 46	16 38	17 29	18 19
20	17 52	18 14	18 40	15 26	16 22	17 18	18 13
30	17 45	18 09	18 37	15 04	16 05	17 05	18 06
35	17 42	18 07	18 36	14 51	15 54	16 58	18 02
40	17 38	18 05	18 36	14 36	15 42	16 50	17 57
45	17 33	18 02	18 36	14 17	15 28	16 40	17 52
S 50	17 27	18 00	18 37	13 54	15 11	16 28	17 46
52	17 25	17 59	18 38	13 43	15 02	16 23	17 43
54	17 22	17 57	18 38	13 31	14 53	16 16	17 40
56	17 19	17 56	18 39	13 17	14 43	16 10	17 36
58	17 15	17 55	18 40	13 00	14 31	16 02	17 32
S 60	17 11	17 53	18 42	12 39	14 17	15 53	17 27

Day	SUN Eqn. of Time 00 h	SUN Eqn. of Time 12 h	SUN Mer. Pass.	MOON Mer. Pass. Upper	MOON Mer. Pass. Lower	Age	Phase
	m s	m s	h m	h m	h m	d	
1	0013	0003	12 00	09 52	22 16	27	●
2	00 06	00 16	12 00	10 41	23 04	28	
3	00 25	00 35	11 59	11 28	23 51	29	

FIGURE 3.4 ☆ *NAUTICAL ALMANAC*, RIGHT HAND DAILY PAGE (TYPICAL)

columns headed: Corr. Record the correction on the worksheet in the space: Corr. ±: 0.0′. Circle the − sign as previously discussed. Since the correction is 0.0′, there is nothing to subtract from Dec. Hours. Therefore, Dec. Hours are rounded off to the nearest minute and recorded in the space headed: True Dec.: 8°11′ N.

Go to the Ass. Position LHA (Almanac) block on your sheet. The next two spaces are not used for sun sights. Draw a slash through these spaces.

Before continuing you will need to familiarize yourself with some techniques for handling longitude:

1. When in West Longitude (which includes all American waters), subtract the *assumed longitude* (Ass. Long.) from the GHA of each sight.
2. When in East Longitude (Fiji, Indian Ocean, etc.), add the assumed longitude to the GHA of each sight.

We are sailing in West Longitude and it will be necessary to subtract the Ass. Long. from Total GHA. But first you must determine whether it is possible to make the subtraction without obtaining a negative number. Note that the Ass. Long., derived from the D.R. Long., is greater than the Total GHA of 105°04′ (rounded to the nearest minute). So, you will need to play a harmless little game to get some degrees to subtract from.

These degrees are easily provided by adding nothing, 0° to the Total GHA except that you add 0° in a special way. Add 360° (full circle) to Total GHA. Record: 360° in the space provided on the worksheet. The sum is rounded off to the nearest minute and recorded on the worksheet in the space: Total GHA: 465°04′.

Subtract the Ass. Long. from Total GHA. . . but, wait a minute, the D.R. Long. is 122°43′ W, not 123°04′ as shown on the form. What's going on? The answer is that you have to obtain the *local hour angle* (LHA) in units of whole degrees without minutes.

If minutes of LHA were used, the sight reduction tables would be so large that we would have to rig a tackle over the boom to lower them into the cabin . . . no, thanks.

However, the value of the Ass. Long. is not arbitrarily selected. The D.R. Long. of 122°43′ W was somewhere between 122°04′ and 123°04′. Select the assumed longitude as close to the dead reckoning longitude as possible. The difference between 123°04′ and 122°43′ was only 21′ whereas, the difference between 122°04′

and 122°43′ was 39′. Therefore, you would select 123°04′ for the assumed longitude.

Subtract Ass. Long. from Total GHA obtaining and recording: LHA: 342°.

Go to the INTERCEPT AZIMUTH—FIRST ENTRY, to the first line, and record: LHA: 342.

Go to the last section of the white pages in the almanac to the tables headed: SIGHT REDUCTION TABLE. These tables will be used to complete the remainder of the worksheet.

The first task will be to select a table corresponding to the *assumed latitude*. Not unlike LHA, the Ass. Lat. must be in degrees without minutes. Since the D.R. Lat. is 34°53′, the nearest degree would be 35°. Record: Lat: 35.

Leafing through the tables we stop at the page headed: LATITUDE/A: 30°–35°.

On the extreme left and right sides of the table are columns headed: LHA/F and LHA respectively. Locate LHA: 342° in the right hand column and go across to the left until you come to the column headed: 35°. The 35° column is the assumed latitude to the nearest degree.

Contained in the latitude column and LHA line intersection are three figures headed: A/H, B/P, and Z_1/Z_2. These figures are required to complete the INTERCEPT AZIMUTH—FIRST ENTRY.

The first figure is recorded in the space on the worksheet labelled: A: 14°40′. If the minutes portion of A were less than 30′, A° would be recorded as 14°. Since A is 14°40′, A° is rounded to the nearest degree and recorded: A°: 15 and A′: 40, as shown.

The second figure is B and is recorded in the space on the worksheet labelled: B±: 53°38′. The sign for B is determined as follows: If LHA is greater than 90° and less than 270°, B is negative (−), and the − sign is circled. If LHA is any other value, B is positive (+) and the + sign is circled. B is positive because LHA is 342° and therefore, greater than 270°, and the + sign is circled. This statement is shown in mathematical language in the upper left hand corner of the table.

The third figure is Z_1 and recorded in the space labelled: Z_1 ±. If B is positive, than Z_1 is positive and the + sign is circled. If B is negative, than Z_1 is negative and the − sign is circled. Since B is positive, Z_1 is recorded: Z_1±: 79.4° and the + sign is circled. This conditional statement is shown in the upper right hand corner of the table.

The next space on the worksheet is labelled: Dec. ±. Move the value of True Dec. to this space and record: Dec. ±: 8° 11′. To determine the + or − sign, you ask yourself the following question? Am I sailing North or South of the equator? If you are sailing North of the equator and the declination is North, the + sign is circled (same name). If you are sailing North of the equator and the declination is South, the − sign is circled (contrary name). Conversely, if you are sailing South of the equator and the declination is South, the + sign is circled (same name). If you are sailing South of the equator and the declination is North, the − sign is circled (contrary name). In our case the + sign is circled because latitude and declination are both North (same name). This condition is stated in the upper left hand corner of the table.

Since B and Dec. are both positive, they are added together, and the Sum of B + Dec is recorded: F: 61°49′. F° is taken to the nearest degree and recorded: F°: 62 and F′: 49.

Armed with all this information we proceed to the next block headed: INTERCEPT AZIMUTH—SECOND ENTRY and record: A°: 15 and F°: 62. Using A° and F° we enter the table for the second time deftly thumbing through to the page headed: LATITUDE/A 12°–17°.

This time we will be using the column headed: 15°, corresponding to A°, and go down the left hand column to 62°, corresponding to F°. At the intersection, the first figure is recorded: H: 58°31′, the second figure is recorded: P°: 60 (rounded to the nearest degree), and the third figure is recorded: Z_2±: 64.0°. If F is greater than 90° then Z_2 is negative and the − sign is circled. If F is less than 90° than Z_2 is positive and the + sign is circled. Since F is less than 90° circle the + sign. This condition is stated mathematically in the upper right hand corner of the table.

Proceed to the next line and record: F′: 49, P°: 60, and Z_2°: 64.

Turn to the last two pages of the tables headed: AUXILIARY TABLE. We will require two corrections from this table.

The first correction is obtained by going across the upper line labelled: F′/A′ to F′: 49 and going down the left hand column labelled: P° to P°: 60. The figure at the intersection is recorded: $Corr_1$ ±: 10′. At the top of the page in the upper left hand corner are two statements: If F is less than 90° and F′ is greater than 29′, then the correction is negative or if F is greater than 90° and F′ is less than 30′, then the correction is also negative. If F degrees and minutes are anything else the correction is positive. You've come a long way. Don't let this confuse you. Let's look at this situation.

F is 61°49′. This means that F is less than 90° and F′ is greater than 29′. This satisfies the first condition and the − sign is circled.

Subtract $Corr_1$ from H obtaining and recording: Sum: 58°21′.

It's time now to re-enter the table for the second correction. This time record: A′: 40 and Z_2°: 64. Go across the top of the table stopping at 40′ and down the right hand column labelled: Z_2° to Z_2: 64. The figure at the intersection is recorded: $Corr_2$ ±: 09′. The sign is negative and the − sign is circled if A′ is less than 30′. Since A′ is greater than 30′ the + sign is circled. This condition is stated mathematically in the upper right hand corner of the table. Adding the Sum: 58°21′ to 09′ we obtain and record: True Hc: 58°30′.

Record: Z_1: 79,4° and Z_2: 64.0°. Adding algebraically and rounding off to the nearest degree, we obtain and record: Z: 143°.

To obtain Zn, four conditions are considered as follows:

1. If sailing in North Latitude and LHA is greater than 180°, then Zn = Z.
2. If sailing in North Latitude and LHA is less than 180°, then Zn = 360° − Z.
3. If sailing in South Latitude and LHA is greater than 180°, then Zn = 180° − Z.
4. If sailing in South Latitude and LHA is less than 180°, then Zn = 180° + Z.

Since we are sailing in North Latitude and our LHA is greater than 180°: Zn = Z and is recorded: Zn: 143°.

Move Ho: 59°00′ down to the True Ho space.

Subtract True Ho from True Hc or True Hc from True Ho to obtain a positive remainder called the *intercept*. In this situation, subtract: True Hc: 58°30′ from: True Ho: 59°00′ obtaining and recording in the space: Intercept: 30′.

Use the following useful memory aid to determine whether the intercept is toward or away from the azimuth of the assumed position. Do not be concerned with understanding the theory at this point. It is more important to learn the procedures. Your questions will all be answered as you progress. The memory aid is "GOAT" and stands for "greater observed angle toward". The observed angle, True Ho, is greater than the calculated angle, True Hc. Check the space labelled: T. If True Hc were greater than True Ho you would have checked the space labelled: A.

The only information still required on the form will be that necessary plot the line of position. The information required to complete the PLOTTING has been determined. Ass. Long. from above is 123°04′ W, Zn is 143° from the middle and below, Ass. Lat. is 35° N from the page used in the sight reduction, the Intercept is 30′ T and the direction toward, from the bottom of the center column.

Note that intercept is shown as minutes of arc on the form. In plotting, intercept distance is expressed as nautical miles. In this case the intercept of 30′ equals 30 nautical miles from the assumed position toward the observed body.

MORRO ROCK, MORRO BAY, CA

FOUR

THE RUNNING FIX

With the worksheet completed, you will want to plot the resultant line of position (LOP).

Begin with the plotting sheet that was used to plot the course from Morro Bay to the DR position on 9/1/86. All information required to plot the line of position has been recorded in the Plotting block on the worksheet.

The plot will be from Ass. Lat. 35° N and Ass. Long. 123°04′ W. Your first task will be to establish this position, called the *assumed position*, on the plotting sheet as shown in Figure 4.1.

Using the MID-LATITUDE and LONGITUDE SCALE, tick off 4′ with the dividers at 35° and transfer the 4′ just left of the 123° meridian. Circle the point and label it AP 1 to represent the assumed position of the first LOP.

Plot the true azimuth, Zn: 143°. Using the true compass, walk 143° over to AP 1 with any parallel plotting device. Because the intercept is toward the 143° draw a dotted line in the direction of the azimuth. With the dividers, measure the distance of 30′ from the vertical scale and draw the azimuth 30′ long.

If the intercept was in the away direction, you would have plotted the azimuth from the assumed position in the opposite direction, that is, 323°.

The appropriate convention is to draw a dashed line for the azimuth.

Draw a solid line of generous length, perpendicular to the azimuth you have established. Label this line 1100 on the top, for local time, and SUN on the bottom, for the body, as shown. This is the proper way to label a line of position.

When the sight was taken, we would have been sailing somewhere on LOP No. 1. That is, if the sight for LOP No. 1 was taken accurately, errors were not made in reducing the sight and the resulting plot of the line of position was accurate.

We do not know what our location is on LOP No. 1 because we need at least one more LOP to provide an intersection. This intersection, provided both sights were handled accurately, would be our terrestrial position.

A third LOP would make us more comfortable (or uncomfortable, as the case may be), dependent upon the size of the triangle formed by the intersection of all three lines of position. If the triangle formed by these three lines were small, we would be confident that our position was well established. If the triangle were large we might take a fourth sight. I have been successful with this method, and usually able to determine the sight to be discarded.

A better method would be to take several sights very close together for each line of position. This technique will be discussed in a later chapter.

Two more sights of the sun were taken over the next two and a half hours on 9/1/86. Complete two additional worksheets by recording the information as follows:

No.: 2; Area: CENTRAL CALIFORNIA COAST; Body: SUN; Date: 9/1/86; Log: 102; WWV: 20h 17m; Stop Watch: 2m 53s;

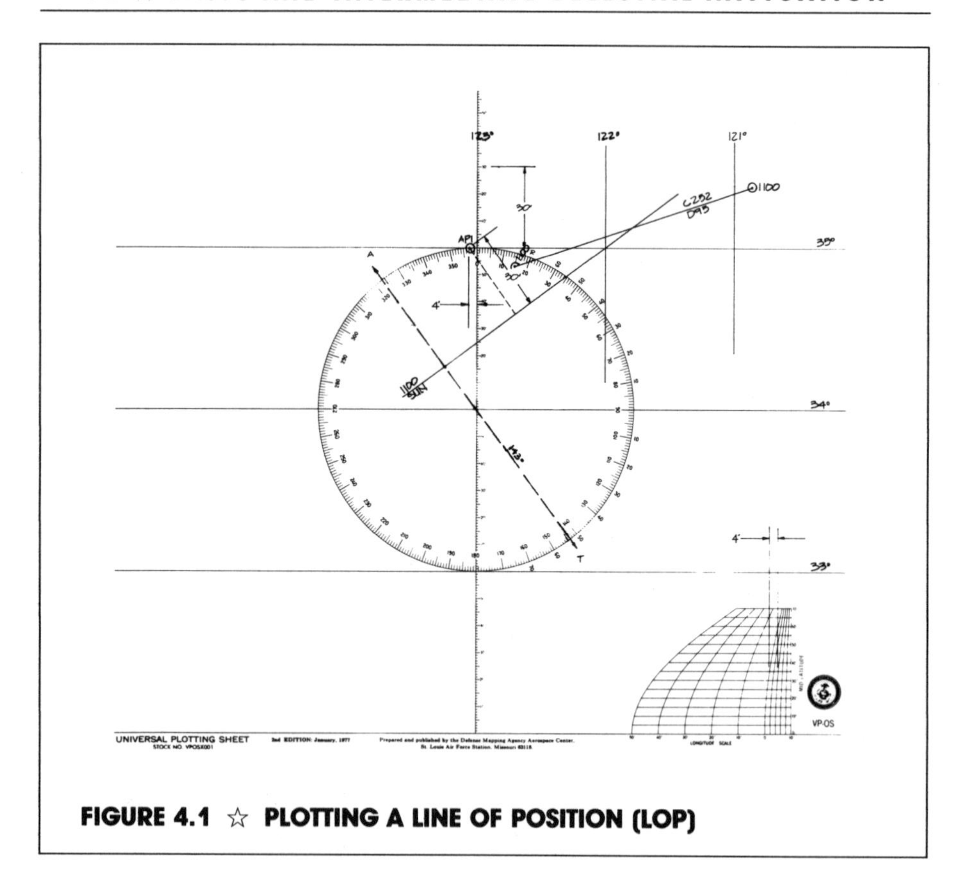

FIGURE 4.1 ☆ PLOTTING A LINE OF POSITION (LOP)

Local Time: 1220; my eye was 12 feet above the water and Hs: 63°17′.

No.: 3; Area: CENTRAL CALIFORNIA COAST; Body: SUN; Date: 9/1/86; Log: 110; WWV: 21h 28m; Stop Watch: 1m 5s; Local Time: 1330; my eye was 12 feet above the water and Hs; 57°58′.

With this information and your previous experience, extend the course line to the two new positions and establish the accompanying dead reckoning positions on the plotting sheet. Use your almanac or Figure 3.4 and the sight reduction tables to complete the worksheets. Plot the last two lines of position on your plotting sheet and check your solutions with Figures 4.2, 4.3, and 4.4

The only difference in reducing the last two sights was that it was not necessary to add 360° to obtain Total GHA for subtracting

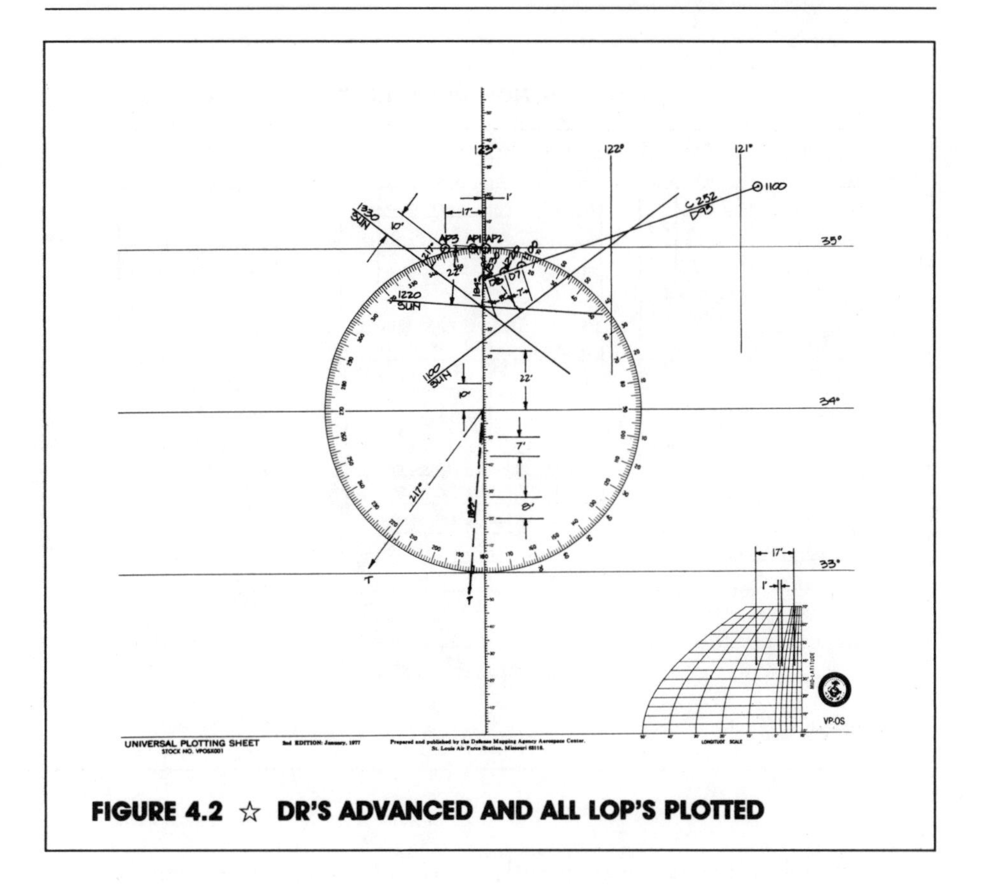

FIGURE 4.2 ☆ DR'S ADVANCED AND ALL LOP'S PLOTTED

the Ass. Long., and LHA was less than 180° so Z was subtracted from 360° to obtain Zn.

These sights were timed so that the lines of position would intersect each other at good angles or "cuts." Experience will help you in the timing of these sights. If the sun is too low in the sky, that is, if the declination is considerably different than your latitude, your sights would be taken over a longer period of time around noon. If the sun is high in the sky, that is, if the declination is close to your latitude, your sights would be taken within a shorter period. An extreme example would be if the sun were of the same declination as your latitude. If this were the case, every sight you took, morning or afternoon, would give you a line of position that would be your longitude. Except for one, very hard, almost impossible

NO. 2

LINE OF POSITION WORKSHEET

AREA CENTRAL CALIFORNIA COAST

BODY	DATE	D. R. LAT.	D. R. LONG.	LOG	WWV	STOP WATCH	LOCAL TIME
SUN	9-1-86	N 34° 51'	W 122° 50'	102	20h 17m	2m 53s	1220

TIME	HR.	MIN.	SEC.
WWV	20	17	—
Stop Watch	—	2	53
GMT	20	19	53

SUN, STARS, & PLANETS SEXTANT (ALMANAC)

Hs	63° 17'	Ha	63° 13.6'
IC	—	Corr.	15.5'
Dip	−3.4'	Add'l Corr.	—
Ha	63° 13.6'	Ho	63° 29'

MOON ONLY SEXTANT (ALMANAC)

Corr.		Hs
Add'l Corr.		IC
HP	Corr.	Dip
Total		Ha
	Total	
	Ho	

ASS. POSITION LHA (ALMANAC)

GHA Hours		120° 00.9'
Min. & Sec.		4° 58.3'
Stars Only SHA		—
Moon & Planets Only V:	Corr.	—
360°		—
Total GHA		124° 59'
Ass. Long.		122° 59'
LHA		2°

INTERCEPT AZIMUTH – FIRST ENTRY

Lat. 35	LHA 2	A 1° 38'	A° 2	A' 38
		B ⊕54° 59'	Z₁ ⊕ 88.9°	
		Dec. ⊕ 8° 10'		
	Sum: B+Dec.	F 63° 09'	F° 63	F' 09

PLOTTING

Ass. Long. W 122° 59'	
Zn 185°	Ass. Lat. N 35°
Intercept	T 22' A

INTERCEPT AZIMUTH – SECOND ENTRY

A° 2	F° 63	H 62° 56'	P° 86	Z₂ ⊕ 86.1°
F' 9	P° 86	Corr.₁ ⊕ 09'	Z₂° 86	
		Sum 63° 05'	Z₁ 88.9°	
A' 38	Z₂' 86	Corr.₂ ⊕ 02'	Z₂ 86.1°	
		True Hc 63° 07'	Z 175°	
		True Ho 63° 29'	Zn 185°	
		Intercept 22'	T ✓ A	

360°
− 175°
185°

DECLINATION (ALMANAC)

Dec. Hours	N 8° 10.1'
'd' ± 0.9	Corr. ± 0.3'
True Dec.	N 8° 10'

FIGURE 4.3 ☆

sight, when the sun was directly overhead, establishing your latitude. I have had this experience, and I usually sight another body such as the moon, star, or planet to obtain my position.

The difference between a *simulaneous fix* and a *running fix* needs to be mentioned. When all sights used to obtain a terrestrial position are taken within a few minutes, such as the sighting of three stars at twilight, and your yacht has not moved more than a fraction of a mile, the intersection is called a fix or simultaneous fix. If two or more sights are taken over a longer period of time and your yacht has moved some distance, the lines of position of the first sights must be advanced along the course line the distance they would have moved at the time of the last sight. The intersection obtained after advancing these lines of position is called a running fix.

A running fix is rarely as accurate as a fix because there are more opportunities for error. Your yacht may not follow the DR course, the log may not be accurate, etc. Don't despair! On most passages you will probably use the sun 95% or more of the time,

NO. 3 **LINE OF POSITION WORKSHEET** AREA CENTRAL CALIFORNIA COAST

BODY	DATE	D. R. LAT.	D. R. LONG.	LOG	WWV	STOP WATCH	LOCAL TIME
SUN	9-1-86	N 34° 48'	W 123°00'	110	21h 28m	1m 5s	1330

TIME

	HR.	MIN.	SEC.
WWV	21	28	—
Stop Watch	—	1	5
GMT	21	29	5

SUN, STARS, & PLANETS SEXTANT (ALMANAC)

Hs	57° 58°	Ha	57° 54.6'
IC	—	Corr.	15.4'
Dip	-3.4'	Add'l Corr.	—
Ha	57° 54.6'	Ho	58° 10'

MOON ONLY SEXTANT (ALMANAC)

Corr.		Hs	
Add'l Corr.		IC	
HP	Corr.	Dip	
Total		Ha	
		Total	
		Ho	

ASS. POSITION LHA (ALMANAC)

GHA Hours		135° 01.1'
Min. & Sec.		9° 16.3'
Stars Only SHA		—
Moon & Planets Only V:	Corr.	—
360°		—
Total GHA		142° 17'
Ass. Long.		123° 17'
LHA		19°

INTERCEPT AZIMUTH – FIRST ENTRY

Lat. 35	LHA 19	A 15° 28'	A° 15	A' 28
		B ⊕53°29'	Z_1 ⊕ 78.8°	
		Dec. ⊕ 8°09'		
	Sum: B+Dec.	F 61° 38'	F° 62	F' 38

PLOTTING

Ass. Long.	W 123° 17'
Zn 217°	Ass. Lat. N 35°
Intercept	T 10' A

INTERCEPT AZIMUTH – SECOND ENTRY

A° 15	F° 62	H 58° 31'	P° 60	Z_2 ⊕ 64.0
F' 38	P° 60	Corr.₁ ± 19'	$Z_2°$ 64	
		Sum 58° 12'	Z_1 78.8°	
A' 28	$Z_2°$ 64	Corr.₂ ± 12'	Z_2 64.0°	
		True Hc 58°00'	Z 143°	
		True Ho 58° 10'	Zn 217°	
		Intercept 10'	T ✓	A

DECLINATION (ALMANAC)

Dec. Hours	N8° 09.2'
'd' ± 0.9	Corr. ± 0.4'
True Dec.	N8° 09'

360°
- 143°
217°

FIGURE 4.4 ☆

and learn to depend on running fixes. Other sights may not be available and a position that is accurate within a few miles is more than acceptable and deeply appreciated.

Back to our little sloop wallowing her way into the enormous North Pacific Ocean. Notice how the plot of our lines of position intersect with a rather large triangle. This is to be expected as the 1100 and 1220 lines of position have not been advanced along the course line to the position they would occupy at 1330. Advance these lines as shown in Figure 4.5.

The 1100 LOP must be advanced 15′ or 15 nautical miles along the course line to the position it would have occupied at 1330. Because the course line is at an angle with the LOP, the actual distance between the 1100 LOP and the 1100 LOP advanced to 1330 is less than 15 nautical miles. It is important to remember to advance lines of position the distance they have moved along the course line only. Replot and relabel the LOP: 1100–1330/SUN and you have recorded this event.

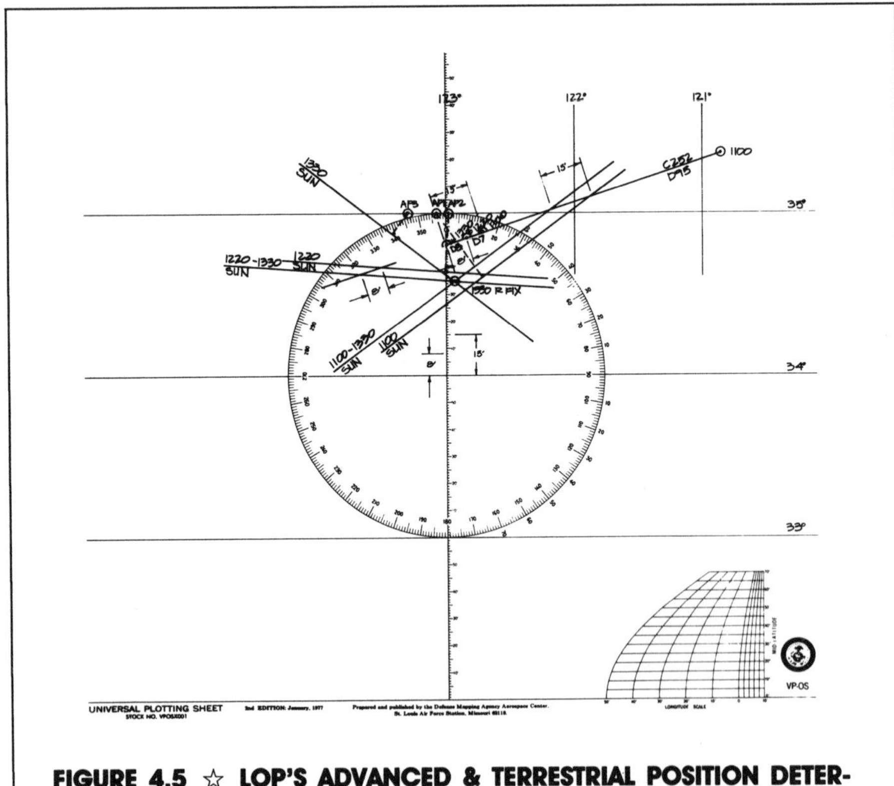

FIGURE 4.5 ☆ LOP'S ADVANCED & TERRESTRIAL POSITION DETERMINED TO BE LATITUDE 34°35′ N AND LONGITUDE 122°56′ W

Advance the 1220 LOP along the course line the distance it would have travelled . . . 8 nautical miles. Note how I extended a small portion of the course line for this purpose. Replot and relabel the LOP: 1220–1330/SUN.

The lines of position, advanced to 1330, form a small triangle. The longest leg of this triangle is approximately 5 nautical miles. Circle a point in the middle of the small triangle and label it horizontally: 1330 R FIX. This is the appropriate convention for labelling a running fix.

With the dividers measure the latitude, and scale the longitude. I ascertained our position to be: 34°35 N 122°56 W. I transferred these coordinates to the North Pacific Ocean chart.

You do not have to transfer these coordinates to a chart. If this

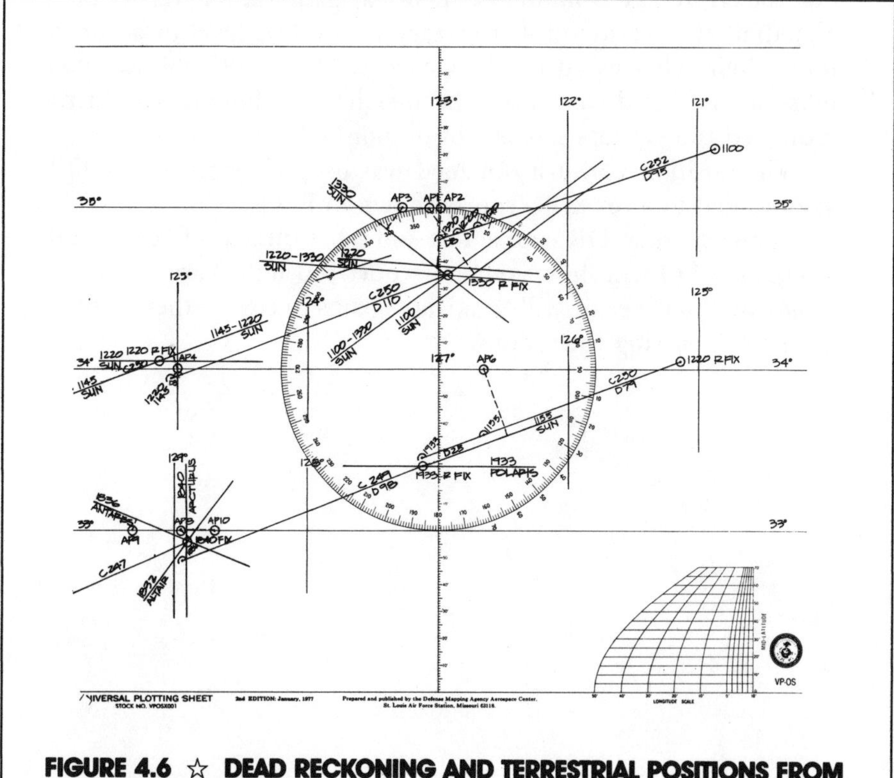

FIGURE 4.6 ☆ DEAD RECKONING AND TERRESTRIAL POSITIONS FROM 9/1/86 TO 9/4/86

were your passage, however, you would want to do this. One of the most exciting times of the day was when we could see our progress on "the big chart."

Finding your position is what celestial navigation is about. This is a typical running fix obtained by taking sights from a small craft. The first sight was taken from the stern rail and the remaining sights from the pitching deck near the mast. This accounts for the difference in height of eye recorded for the latter sights.

This level of accuracy would have been frowned on if these sights were obtained from the stable bridge of a large ship. I want you to realize that I am pleased with this determination of our terrestrial position, and that it is about the best that can be expected without spending the entire day on navigation. This is not to say that the entire day wouldn't be spent navigating if we were making a landfall. I have found it necessary to use this level of accuracy for landfalls when extensive cloud cover made it impossible to obtain more sights. Land has usually been sighted within an hour of the estimated time except in cases of diminished visibility.

The position you just obtained was used to begin a new DR. We decided to steer 250° true as shown in Figure 4.6.

Draw the new DR course line and label the top of line: C250. In Chapter 12 I will show you how we arrived at the various courses or *sailings;* however, it will be advantageous to cover other material prior to discussing the sailings.

SPERM WHALE, NORTH PACIFIC

FIVE

THE COURSE LINE

There is a technique I have found useful while passage-making. The technique involves taking a timed sight of the sun such that the resulting line of position, called a *course line*, is plotted parallel with the DR course. The parallel LOP serves three purposes.

1. The course line LOP does not have to be redrawn when advanced to a later time along the DR course.
2. The course line LOP is a visual indication of how far you have sailed above or below the DR course.
3. The course line LOP combined with an LOP from another sight of the sun or other celestial body provides an easily-obtained running fix.

It was our second day at sea and we wanted to begin this system of monitoring our course.

It was necessary to find the exact time when the azimuth of the sun would be perpendicular to the DR course. The method used for finding the exact time to sight the sun requires working backward, and several estimates of the time might be required.

Begin with a Line of Position Worksheet and record as follows: No. 4; Area: NORTH PACIFIC; Body: SUN; and Date: 9/2/86 as shown in Figure 5.1.

I began to complete the worksheet at about 0900 hours local time. However, do not record this time as it is not the time of the sight.

The worksheet is designed with space for calculations in the lower right hand corner. Record the course: 250° as shown. Subtract 90° from the 250° to obtain and record the desired azimuth of the sun: 160°.

You will be making an approximation of the dead reckoning position and determining when the azimuth of the sun will be perpendicular to the course at the predetermined position. Think of the azimuth as nothing more than an arrow that points to the sun. If you were sailing north of the sun's declination, the arrow would point in a direction southeast of your position in the morning, directly south or 180° at 1200 local apparent time (which almost

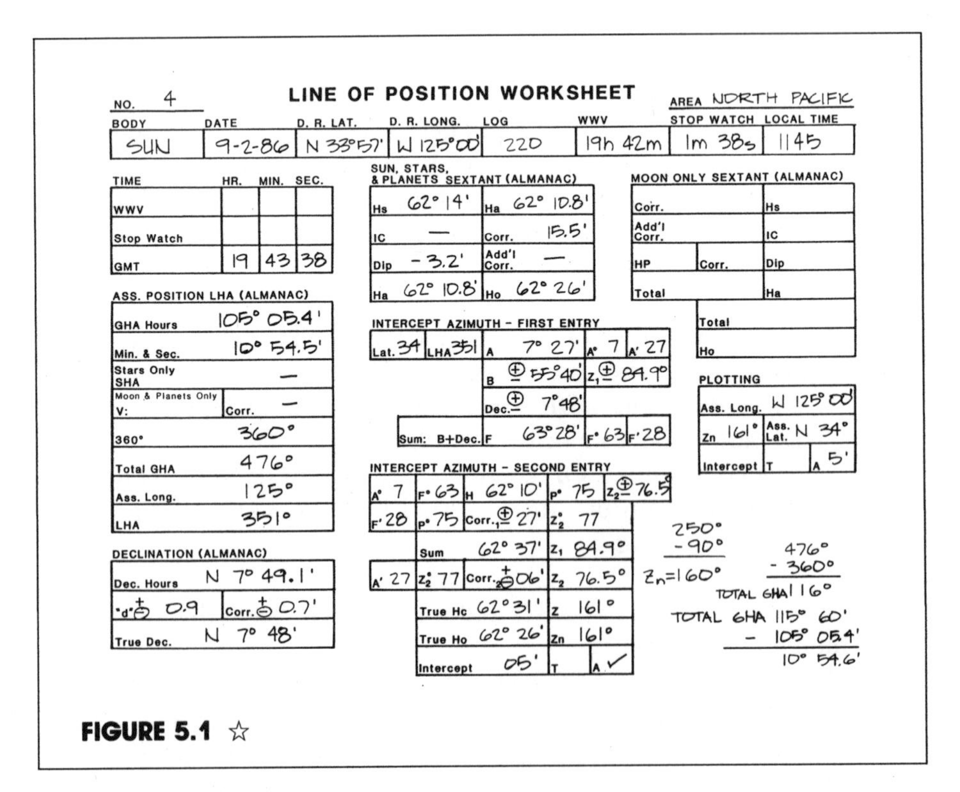

LINE OF POSITION WORKSHEET

NO. 4 — AREA NORTH PACIFIC

BODY	DATE	D. R. LAT.	D. R. LONG.	LOG	WWV	STOP WATCH	LOCAL TIME
SUN	9-2-86	N 33°57'	W 125°00'	220	19h 42m	1m 38s	1145

TIME	HR.	MIN.	SEC.
WWV			
Stop Watch			
GMT	19	43	38

ASS. POSITION LHA (ALMANAC)

GHA Hours	105° 05.4'
Min. & Sec.	10° 54.5'
Stars Only SHA	—
Moon & Planets Only V: / Corr.	—
360°	360°
Total GHA	476°
Ass. Long.	125°
LHA	351°

DECLINATION (ALMANAC)

Dec. Hours	N 7° 49.1'
'd' ± 0.9	Corr. ± 0.7'
True Dec.	N 7° 48'

SUN, STARS, & PLANETS SEXTANT (ALMANAC)

Hs	62° 14'	Ha	62° 10.8'
IC	—	Corr.	15.5'
Dip	− 3.2'	Add'l Corr.	—
Ha	62° 10.8'	Ho	62° 26'

INTERCEPT AZIMUTH – FIRST ENTRY

Lat. 34 | LHA 351 | A 7° 27' | A° 7 | A' 27

B ⊕ 55°40' | Z₁ ⊕ 84.9°

Dec. ⊕ 7°48'

Sum: B+Dec. F 63°28' | F° 63 | F' 28

INTERCEPT AZIMUTH – SECOND ENTRY

A° 7 | F° 63 | H 62° 10' | P° 75 | Z₂ ⊕ 76.5

F' 28 | P° 75 | Corr.₁ ⊕ 27' | Z₂ 77

Sum 62° 37' | Z₁ 84.9°

A' 27 | Z₂ 77 | Corr.₂ ± 06' | Z₂ 76.5°

True Hc 62°31' | Z 161°

True Ho 62° 26' | Zn 161°

Intercept 05' | T | A ✓

MOON ONLY SEXTANT (ALMANAC)

Corr.		Hs
Add'l Corr.		IC
HP	Corr.	Dip
Total		Ha
Total		
Ho		

PLOTTING

Ass. Long.	W 125° 00'
Zn 161°	Ass. Lat. N 34°
Intercept T	A 5'

250°
− 90°
Zn = 160°

476°
− 360°
TOTAL GHA 116°

TOTAL GHA 115° 60'
− 105° 05.4'
10° 54.6'

FIGURE 5.1 ☆

always will not be the same as 1200 clock time), and somewhere southwest of your position in the afternoon.

Consult the lines of position at 1100–1330/SUN and 1220–1330/SUN as plotted on the previous day. See how the C250 DR course nearly bisects these two LOPs. It is reasonable that the course line LOP would have a perpendicular azimuth at somewhere between 1135 and 1150 hours. I estimated the time at approximately 1145.

Knowing our progress, the DR course line was advanced to a position we expected to occupy at 1145 local time. Record the position: D.R. Lat.: 33°57′ N, D.R. Long.: 125°00′ W, and Log: 220 (nautical miles).

Plot the 1145 DR position as shown in Figure 4.6, and label the distance sailed: D 110 under the course line as shown.

Record: Local Time: 1145. When the log was checked at 1145 it read 221 . . . only one mile further than anticipated.

D.R. Long. will be used for Ass. Long. Record the assumed longitude in the spaces provided in the blocks entitled: Ass. Position LHA (Almanac) and Plotting: Ass. Long.: 125° W. Record: Ass. Lat.: 34° N in the space provided in the plotting block.

Open the almanac to the right hand white page headed: 1986 SEPTEMBER 1, 2, 3, (MON., TUES., WED.) or go to Figure 3.4.

Go down the G.M.T. column to: TUESDAY at 19 hours, across to the columns headed: SUN and stop at the column headed: G.H.A. Obtain and record: GHA Hours: 105°05.4′. Go across to the column headed: Dec. and record: Dec. Hours: 7°49.1′ N. Go to the bottom of the page and record: "d" ±: 0.9 and circle the minus sign because declination is decreasing.

Go to the yellow page headed: 45m and for the first approximation of the time use 45m 00s corresponding to 11°15.0′ from the column headed: SUN PLANETS. Adding 105°05.4′ and 11°15.0′ (on a separate piece of paper) we obtain: 116°20.4′. Add to this the figure 360° obtaining : 476°20.4′. Drop the 20.4′. That is 476° will be our first approximation of our Total GHA. Subtract the Ass. Long.: 125° from 476° obtaining LHA: 351°.

Obtain the correction for the declination and record: Corr: 0.7′ while circling the − sign. Therefore, True Dec.: 7°48′ N is recorded.

Turn to the sight reduction table and stop on the page headed: LATITUDE/A: 30°–35°. Go down the column headed: 34°, stop at the intersection corresponding to LHA: 351°, and begin the sight reduction procedure as in previous examples.

After recording Z_2 add Z_1 and Z_2 obtaining: 161.4°. Since Zn is very close to 160° the value Zn = 161° is a very close approximation, and another entry will not be required. If there was a difference of over 2° in azimuth I would have increased or decreased my approximation of the time of the sight and recalculated the LHA until the appropriate value of Zn was selected. You also know from your previous sights that LHA is greater than 270° and that you are in North Latitude, so Zn = Z.

Complete the altitude calculation and check your results with Figure 5.1.

Go to the lower right hand portion of the worksheet, subtract 360° from 476° and obtain the Total GHA that is less than the assumed longitude. Rewrite Total GHA as 115°60′.

Returning to the almanac, we see that the largest value of GHA that can be subtracted corresponds with 19h GMT and is 105°05.4′. Subtract 105°05.4′ from 115°60′, obtaining an arc of 10°54.6′.

Go to the yellow pages headed: INCREMENTS AND CORRECTIONS to find the corresponding minutes and seconds of the arc 10°54.6′. Go to the half of the yellow page headed: 43m., go down the column headed: SUN/PLANET to 10°54.5′, the closest arc obtainable. Go to the left to 38s. Record in the Time block spaces next to GMT and under the headings: Min.: 43 and Sec.: 38. The exact time for taking the course line sight has been determined.

For reference, record in the space: Min. & Sec.: 10°54.5′.

All that remains is to time the sun sight.

The sextant was checked for index error and none was found.

The stopwatch was started and GMT was recorded: WWV: 19h 42m. 1m 38s was left to obtain the sight. My wife counted down so that our sight would be exactly on time.

At 1m 38s later I brought the lower limb of the sun to the horizon, the stopwatch was stopped, and the sextant was read and recorded: Hs: 62°14′.

The sight was corrected for the height of my eye, in this case 11 feet.

Ho was figured to be 62°26′ and entered in the True Ho box for determination of the intercept.

Plot and label the course line LOP: 1145/SUN. Note that the course line LOP appears to be approximately 8 nautical miles above the DR course line plotted from yesterday's position. If the sight was taken accurately, the 8-mile difference would appear to be reasonable.

Many unpredictable factors can cause a displacement of the intended DR course. We were just beginning our passage and were "zeroing in" on our course. Note we were approximately 10 miles below our DR when we determined our terrestrial position yesterday.

The few extra minutes required to reduce a course line sight are more than compensated for by the information gained by the resultant LOP. On longer passages, where there are no permanent obstructions and only moderate accuracy is necessary, the course line sight and one other sight will yield a reasonable terrestrial position from day to day.

HOKULEA, HONOLULU, HI

SIX

THE NOON SIGHT

The noon sight is older than the discovery of the New World, and a book on celestial navigation would not be complete without some instruction and a discussion of the subject.

In fact, celestial navigation was thought to be analogous with magic, was a protected art prior to modern times, and was taught to masters and mates only. Seamen, exposed to the worst conditions, poorly treated, and often "Shanghaied" from the ports of the world, could have mutinied if they knew how to navigate. Recall the famous story of the mutiny on the HMS *Bounty*. The only way the crew could successfully take over the Bounty was by convincing the first mate, a navigator, to throw in with them.

If you were to observe the art of celestial navigation in those days it might have looked like this. . . . The captain, the mate and the cabin boy would appear on the poop deck with sextant, hourglass and other deceptive devices at sometime prior to the *meridian passage* of the sun.

They would perform various rituals from different positions on the poop and appear to be making many different types of observations. The cabin boy would occasionally turn over the hour glass and give the appearance of timing these observations. The captain or the mate would occasionally check the progress of the sun, pass the sextant back and forth to each other, and take *apparent* sights of different phenomena from different locations. Then they would observe the sun as it rose to its highest position in the sky at meridian passage.

The sun was all they were really concerned with. Amongst all this activity they would record the sun sight at *local apparent noon* (LAN) while continuing with their sham, escaping detection from the crew.

It is important to mention that they would allow plenty of time prior to LAN to follow the progress of the sun to its highest altitude.

The officer would make some corrections to the altitude and apply the declination to obtain the latitude.

These early navigators were unable to keep time because the most accurate timepieces were constructed with massive pendulums and could not operate at sea. Without a way to tell time it was not possible to obtain the longitude, so the ship would sail a course to the desired latitude, short of its destination, then "sail down the latitude" until land was sighted. They had no way of knowing when land would be sighted. This procedure created a lot of stress and many a ship's master was greatly relieved when the announcement: "land ho," was heard from the crow's nest.

As you know, we were into our second day at sea and my wife, my only crew member, was too seasick to mutiny. So without deception I decided to obtain a sight at our local noon.

Begin by extending the DR course line another three miles as shown in Figure 4.6. The extra 3 nm establishes our assumed position for the sight at LAN.

Record the following information on a worksheet: No.: 5; Area: NORTH PACIFIC; Body: SUN; Date: 9/2/86; D. R. Long.: 125°04′

W, Lat.: 33°56′ N; Log: 223, and Local Time: 1220, as shown in Figure 6.1. The 1220 time was my best estimate of when the sun would be due south of my position.

Go to the lower right hand corner of the worksheet and record the D. R. Long.: 125°04′ W.

Open the almanac to the right hand page for 9/2/86 or Figure 3.4.

Go down the first column headed: G.M.T. to TUESDAY the 2nd and the second column headed: SUN, G.H.A. to find the GMT hour that is the closest to and a little less than the D. R. Longitude. At 20h GMT the GHA of the sun is 120°05.6′. Record in the Time block and in the calculation area of the worksheet: HR: 20h = 120°05.6′ respectively.

While in the almanac, and on the same line, obtain and record: Dec. Hours: N 7°48.2′; "d" ±: 0.9 and circle the − sign.

Subtract the GHA at 20h GMT from D. R. Longitude as shown and obtain a difference of 4°58.4′.

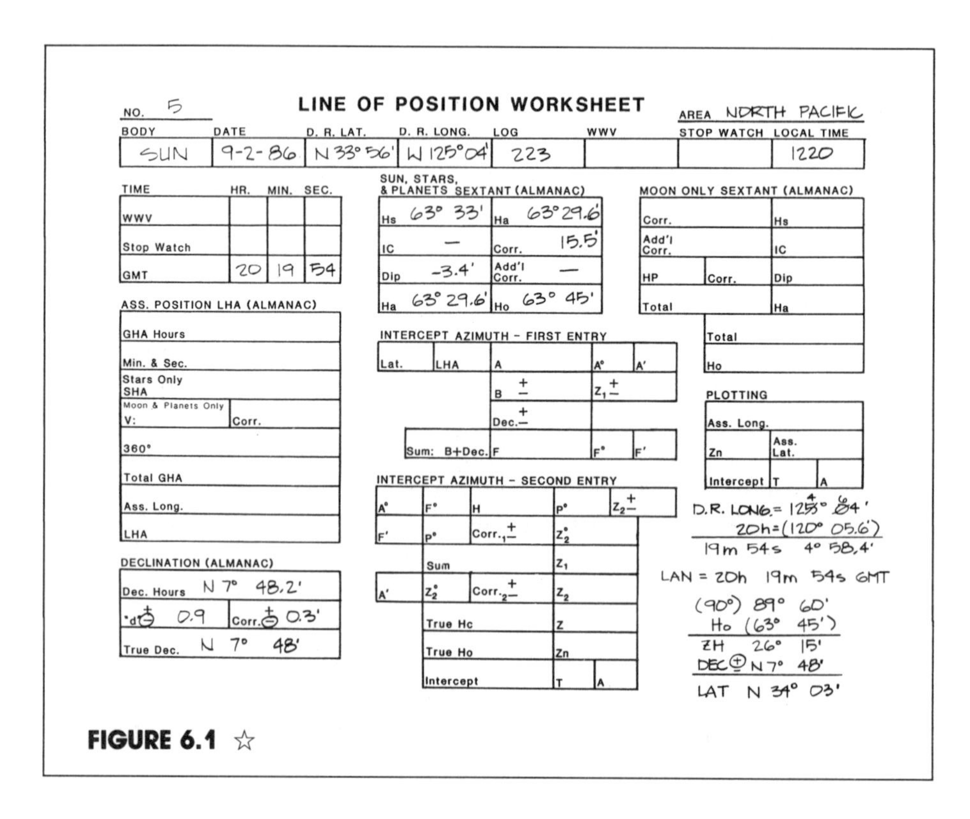
NO. 5 LINE OF POSITION WORKSHEET AREA NORTH PACIFIC

BODY	DATE	D. R. LAT.	D. R. LONG.	LOG	WWV	STOP WATCH	LOCAL TIME
SUN	9-2-86	N 33° 56′	W 125° 04′	223			1220

TIME	HR.	MIN.	SEC.
WWV			
Stop Watch			
GMT	20	19	54

SUN, STARS, & PLANETS SEXTANT (ALMANAC)

Hs	63° 33′	Ha	63° 29.6′
IC	—	Corr.	15.5′
Dip	−3.4′	Add'l Corr.	—
Ha	63° 29.6′	Ho	63° 45′

MOON ONLY SEXTANT (ALMANAC)

Corr.		Hs
Add'l Corr.		IC
HP	Corr.	Dip
Total		Ha
	Total	
	Ho	

ASS. POSITION LHA (ALMANAC)

GHA Hours	
Min. & Sec.	
Stars Only SHA	
Moon & Planets Only V:	Corr.
360°	
Total GHA	
Ass. Long.	
LHA	

INTERCEPT AZIMUTH – FIRST ENTRY

Lat.	LHA	A	A°	A′
		B ±	Z₁ ±	
		Dec. ±		
Sum: B+Dec.		F	F°	F′

PLOTTING

Ass. Long.		
Zn	Ass. Lat.	
Intercept	T	A

INTERCEPT AZIMUTH – SECOND ENTRY

A°	F°	H	P°	Z₂ ±
F′	P°	Corr.₁ ±	Z₂°	
	Sum		Z₁	
A′	Z₂°	Corr.₂ ±	Z₂	
	True Hc		Z	
	True Ho		Zn	
	Intercept		T	A

DECLINATION (ALMANAC)

Dec. Hours N 7° 48.2′	
"d" ± 0.9	Corr. ± 0.3′
True Dec. N 7° 48′	

D.R. LONG. = 125° 04′
20h = (120° 05.6′)
19m 54s 4° 58.4′
LAN = 20h 19m 54s GMT
(90°) 89° 60′
Ho (63° 45′)
ZH 26° 15′
DEC ⊕ N 7° 48′
LAT N 34° 03′

FIGURE 6.1 ☆

Using the remainder: 4°58.4′, open the almanac to the yellow pages headed: INCREMENTS AND CORRECTIONS.

Go to the column headed: SUN/PLANETS and go through the pages until you find the increment of GHA closest to the remainder. The closest value is 4°58.5′. Record in the Time block spaces and on the calculation section: Min.: 19m and Sec.: 54s and 19m 54s respectively, obtaining local apparent noon (LAN) at our assumed position.

Obtain and record the correction for little d: Corr. ±: 0.3′ and circle the − sign. Subtract the 0.3′ and round off the answer, obtaining: True Dec.: 7°48′ N.

I began listening to WWV and WWVH at about 20h 15m.

I was attached to the port side after lower shroud via my safety harness, and I sat on the cabin top facing the sun.

The sextant had no index error, and after finding a comfortable index shade I began to observe the sun. The technique used in bringing the lower limb of the sun to the horizon is the same as previously discussed. Once the sun is on the horizon, however, the procedure is to continue to follow the sun to its maximum altitude while slowly turning the micrometer drum in one direction only.

When the maximum altitude is reached the sun will appear to hang there for awhile. There is a tendency to turn the micrometer drum in the opposite direction; however, don't do it. By avoiding this tendency and not changing the settings on the sextant, you have measured the altitude at LAN. In this case the sextant altitude was recorded: Hs: 63°33′.

We did not time the sight.

My eye was approximately 12′ above the water. Dip and lower limb altitude corrections were applied and rounded to obtain Ho: 63°45′.

To obtain the latitude subtract Ho from 90°, obtaining the *zenith height* (ZH). Rewrite 90° as 89°60′ and subtract: Ho: 63°45: yielding: ZH: 26°15′.

Some rules are applied to determine whether the declination is added or subtracted from the zenith height as follows:

1. If Dec. & Lat. same name N or S, add Dec. to ZH and give Lat. same name.
2. If names differ N or S, subtract Dec. from ZH and name Lat. opposite from Dec.

One other case requires clarification:

3. If sailing at Lat. between Dec. and Equator N or S, subtract ZH from Dec. and remainder is Lat.

In our case Lat. & Dec. are the same name. Lat. and Dec. are added and the Lat. is 34°03′ N.

Plot the latitude as a horizontal line of position and label it 1220/SUN.

As previously mentioned, the nice thing about advancing the earlier course line LOP from 1145 to 1220 is that it is simply relabelled: 1145–1220/SUN and a running fix has been determined. A better fix would result if the LOPs intersected at an angle greater than 30°.

Label the position horizontally: 1220 R FIX and begin the new DR course line. The course is still 250° true and is labelled: C250.

We transferred the coordinates of the position to the North Pacific Ocean chart. We determined: Lat. 34°03′ N and Long. 125°09′ W.

A revolutionary event for celestial navigation occured during the year 1735. John Harrison, clockmaker, delivered the first in a series of large, cumbersome, springwound chronometers designed to keep time at sea, to the Royal Observatory at Greenwich, England. These chronometers were ingenious mechanical devices and weighed about fifty pounds. One of his models was selected to be tested by one of the world's greatest navigators, Captain James Cook, aboard HMS *Resolution*. The trial chronometer operated successfully, thus paving the way for accurate celestial navigation. The chronometer made possible the determination of GMT as required to solve the formulas that are basic to the creation of the tables you have been using. With GMT, it became possible for the first time to determine longitude and accurately chart the seas.

Happily for us, a quartz crystal wristwatch of today's design weighs a few ounces and keeps better time than the 18th century chronometer.

It is possible to obtain the longitude by a series of sightings of the sun around noon. Here is how this is accomplished:

1. Take three sights of the sun before LAN and record the altitudes and times.
2. Take the LAN sight and record its altitude.

3. Take three sights of the sun at the same altitudes as the first three sights after LAN and record the times.
4. Average the times of the six sights for GMT at LAN.
5. Convert GMT to GHA as shown below to obtain the longitude.

See Figure 6.2 and the example below:

Hs(a) & (a′)	= 61°00′	GMT(a)	=	19h 50m 13s
		GMT(a′)	=	20h 50m 42s
Hs(b) & (b′)	= 62°00′	GMT(b)	=	20h 05m 20s
		GMT(b′)	=	20h 35m 32s
Hs(c) & (c′)	= 63°00′	GMT(c)	=	20h 15m 10s
		GMT(c′)	=	20h 25m 35s

Hs(LAN) = 63°33′. The Lat. was determined in the previous example.

GMT(LAN) = {GMT(a) + GMT(a′) + GMT(b) + GMT(b′) + GMT(c) + GMT(c′)}/6

GMT(LAN) = {119h + 180m + 152s}/6 = {120h + 120m + 152s}/6

GMT(LAN) = 20h + 20m + 25.3s = 20h 20m 25s.

Go to the daily page in the 1986 *Nautical Almanac* for September 2 or Figure 3.4 and the yellow pages of your almanac to obtain:

20h GMT	120°05.6′
20m 25s	5°06.3′
	125°11.9′ W = 125°12′ W the longitude.

The above examples are presented as possibilites for finding latitude and longitude. Sight reduction tables are not required, hence these methods could be used in lifeboat navigation. The latitude sight is very accurate, but is not obtainable if a cloud obscures the sun around LAN. Visibility around noon is essential for this kind of navigation and this is a serious limitation. Visibility is not so much of an issue when using sight reduction tables because any combination of sights can be taken during short periods when the sun is visible. The longitude method again is prone to visibility limitations because of the multiplicity of sights required. The multiple sights should be plotted as in Figure 6.2 and a sight, or sights, discarded if not near its expected position. This method can yield errors of magnitude ± 15 nautical miles or more in longitude.

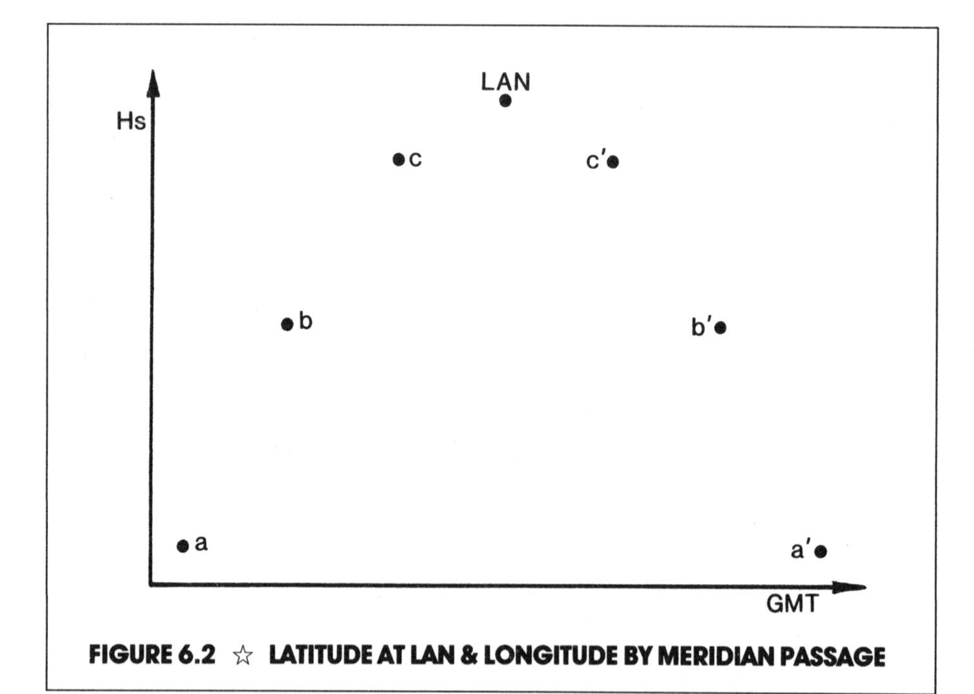

FIGURE 6.2 ☆ LATITUDE AT LAN & LONGITUDE BY MERIDIAN PASSAGE

Sight reduction tables have made those methods outdated. There are good reasons to know the techniques, however. I use the latitude sight as a check when I can get it, and the longitude sight method is presented as a possible survival technique. I made up the longitude portion of the problem because I do not usually take the time to do it.

And now we return to our passage. The following morning, 9/3/86, I decided to determine how we were progressing in reference to the DR course.

Go to Figure 4.6. See how the running fix from the prior day was replotted on the right side of the plotting sheet. A new plotting sheet could have been used, but I want to show you a method of saving paper. An entire passage can be plotted on both sides of two plotting sheets. Depending on the length of your passage, they can become a little cramped so I don't necessarily recommend it. Also do not forget to replot the meridians for the change in latitude by using the scale in the right hand lower corner.

Circle the point and label it 1220 R FIX.

Advance the DR 79 course line nautical miles. The course is 250° true.

Label the DR 1155.

Begin LOP Worksheet No. 6 and determine the time the sight will be taken.

At the determined time Hs was 62°24′, there was no index error, and my eye was 11′ above the water.

Plot the course line sight and label the LOP: 1155/SUN.

Go to Figure 4.6 and Figure 6.3 to check your work.

The timed sight of the sun was approximately six minutes later than the timed sight of the sun for yesterday's course line. The time difference was expected because we were sailing west.

A noon sight could have been taken at approximately six minutes later than the previous day's LAN. You will become aware of the daily local time changes at sea. These details will help you in determining when to do your navigating.

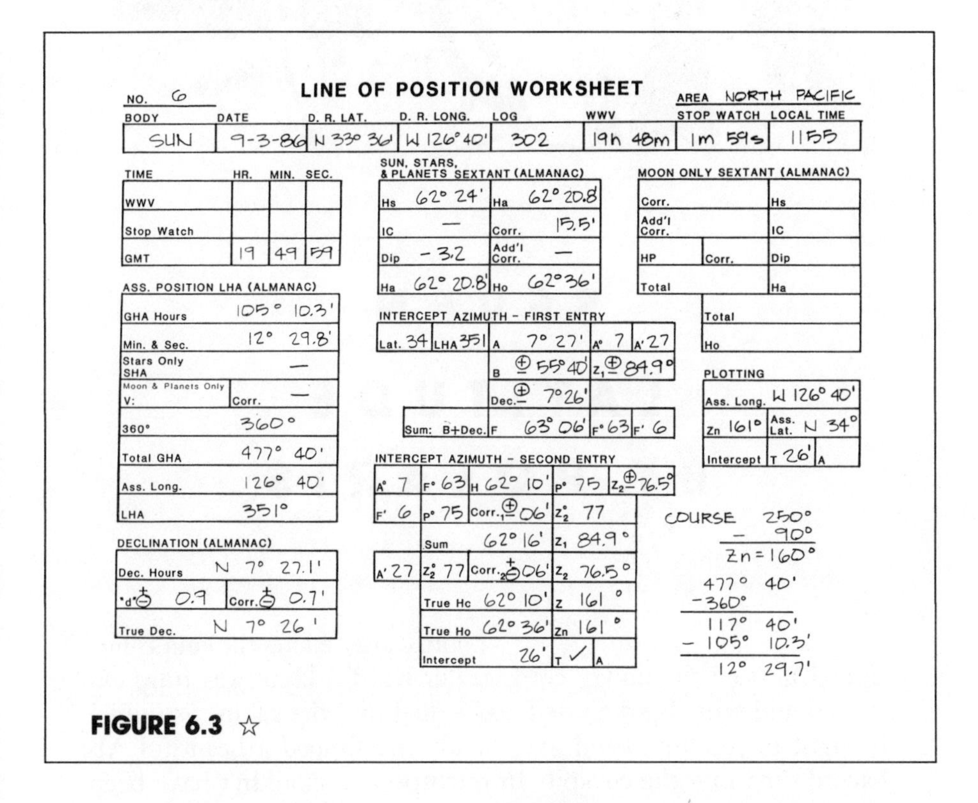

NO. 6 LINE OF POSITION WORKSHEET AREA NORTH PACIFIC

BODY	DATE	D. R. LAT.	D. R. LONG.	LOG	WWV	STOP WATCH	LOCAL TIME
SUN	9-3-86	N 33° 36′	W 126° 40′	302	19h 48m	1m 59s	1155

TIME	HR.	MIN.	SEC.
WWV			
Stop Watch			
GMT	19	49	59

ASS. POSITION LHA (ALMANAC)

GHA Hours	105° 10.3′
Min. & Sec.	12° 29.8′
Stars Only SHA	—
Moon & Planets Only V: / Corr.	—
360°	360°
Total GHA	477° 40′
Ass. Long.	126° 40′
LHA	351°

DECLINATION (ALMANAC)

Dec. Hours	N 7° 27.1′
"d" ± 0.9	Corr. ± 0.7′
True Dec.	N 7° 26′

SUN, STARS, & PLANETS SEXTANT (ALMANAC)

Hs	62° 24′	Ha	62° 20.8′
IC	—	Corr.	15.5′
Dip	−3.2	Add'l Corr.	—
Ha	62° 20.8′	Ho	62° 36′

MOON ONLY SEXTANT (ALMANAC)

Corr.		Hs
Add'l Corr.		IC
HP	Corr.	Dip
Total		Ha
		Total
		Ho

INTERCEPT AZIMUTH – FIRST ENTRY

Lat. 34	LHA 351	A 7° 27′	A° 7	A′ 27
		B ⊕ 55° 40′	Z_1 ⊕ 84.9°	
		Dec. ⊕ 7° 26′		
Sum: B+Dec.		F 63° 06′	F° 63	F′ 6

PLOTTING

Ass. Long.	W 126° 40′
Zn 161°	Ass. Lat. N 34°
Intercept	T 26′ A

INTERCEPT AZIMUTH – SECOND ENTRY

A° 7	F° 63	H 62° 10′	P° 75	Z_2 ⊕ 76.5°
F′ 6	P° 75	$Corr._1$ ⊕ 06′	$Z_2°$ 77	
		Sum 62° 16′	Z_1 84.9°	
A′ 27	$Z_2°$ 77	$Corr._2$ ± 06′	Z_2 76.5°	
		True Hc 62° 10′	Z 161°	
		True Ho 62° 36′	Zn 161°	
		Intercept 26′	T ✓	A

COURSE 250°
− 90°
Zn = 160°

477° 40′
−360°
117° 40′
− 105° 10.3′
12° 29.7′

FIGURE 6.3 ☆

SEVEN

LATITUDE BY POLARIS

Shortly after taking the course line sight I noticed the bungy cord on our hand fishline was fully extended and taut. I got excited and yelled into the cabin, fish on! I brought in the line, hand over hand, and flipped a beautiful Albacore tuna into the cockpit. In retrospect, I shouldn't have been so hasty. I had not put on leather gloves and there are many cases where fingers have been cut badly or lost when a heavy fish decided to run at the wrong time. I estimated the tuna to be approximately thirty pounds. Well . . . at least twenty-five pounds.

After filleting him and enjoying a scrumptious luncheon, we cut him into proper size pieces, sterilized mason jars, added a little water and cooked the little jars full of "Charley" in our stainless steel pressure cooker. It took the greater part of the day to can him. The mason jars all sealed properly and we finished eating the contents of our last jar seven months later. That albacore was the best tasting canned tuna I've ever eaten.

What does this all have to do with sighting Polaris? Everything. I didn't have time during the day on 9/3/86 to take a second sight. Toward evening I decided to determine our position by combining the DR course line sight taken late morning with a sight of the pole star. The wind was light, the sea was moderate, and the sky was clear as twilight approached. To prepare for taking the sight I needed to determine the best working period during the twilight time. The sight would be taken at a time when Polaris and the horizon would both be visible.

Begin by filling in a worksheet as follows: No.: 7; Area: NORTH PACIFIC; Body: POLARIS; and Date: 9/3/86, as shown in Figure 7.1.

Open the almanac to the right hand page dated 1986 SEPTEMBER 1,2,3 (MON., TUES., WED.) or Figure 3.4. The information on the right hand half of the page will be used.

The right hand half of the page is divided into two sections: the upper right hand half: Lat., Twilight, Sunrise, and Moonrise and the lower right hand half: Lat., Sunset, Twilight, and Moonset.

Go to the lower right hand half of the page to the columns headed: Twilight, Civil and Nautical.

Before working with these tables we need to discuss some aspects of time.

I keep the ship's clock on local time or *zone time*. Local time is nothing more than the time we are all accustomed to. The time begins at the prime meridian (Greenwich) and changes by one hour for every 15° of longitude.

So as not to confuse anyone, let's agree that we will keep 24-hour time. In 24 hours the earth rotates one revolution, a total of 360°. Dividing 360° by 24 hours we obtain 15° per hour. Therefore, the earth rotates 15° per hour.

The easiest method of keeping the ship's time is to change the

NO. 7 **LINE OF POSITION WORKSHEET** AREA NORTH PACIFIC

BODY	DATE	D. R. LAT.	D. R. LONG.	LOG	WWV	STOP WATCH	LOCAL TIME
POLARIS	9-3-86	N 33° 27'	W 127° 08'	327	3h 31m	2m 2s	1933

TIME	HR.	MIN.	SEC.
WWV	3	31	–
Stop Watch	–	2	2
GMT	3	33	2

SUN, STARS, & PLANETS SEXTANT (ALMANAC)

Hs	33° 01'	Ha	32° 57.8'
IC	—	Corr.	– 1.5'
Dip	– 3.2'	Add'l Corr.	* 28'
Ha	32° 57.8'	Ho	33° 24'

MOON ONLY SEXTANT (ALMANAC)

Corr.		Hs
Add'l Corr.		IC
HP	Corr.	Dip
Total		Ha
		Total
		Ho

ASS. POSITION LHA (ALMANAC)

GHA Hours	♈ 27° 56.6'
Min. & Sec.	♈ 8° 16.9'
Stars Only SHA	—
Moon & Planets Only V: / Corr.	—
360°	360°
Total GHA	396° 14'
Ass. Long.	127° 14'
LHA	♈ 269°

DECLINATION (ALMANAC)

Dec. Hours	
"d" ±	Corr. ±
True Dec.	

INTERCEPT AZIMUTH – FIRST ENTRY

Lat.	LHA	A	A°	A'
		B ±	Z_1 ±	
		Dec. ±		
	Sum: B+Dec.	F	F°	F'

INTERCEPT AZIMUTH – SECOND ENTRY

A°	F°	H	P°	Z_2 ±
F'	P°	Corr.₁ ±	Z_2°	
	Sum		Z_1	
A'	Z_2°	Corr.₂ ±	Z_2	
	True Hc		Z	
	True Ho		Zn	
	Intercept		T	A

PLOTTING

Ass. Long.	
Zn	Ass. Lat.
Intercept T	A

LAT = APP ALT. – 1° + A_0 + A_1 + A_2

A_0 = 1° 26.6'
A_1 = 0.5'
A_2 = 0.9'
1° 28'
– 1°
28'
*ADD'L CORR.

SUNSET
1919 TW NAUT.
(1850) TW CIVIL
2 129 WORKING
15 PERIOD
1850
1905 LOCAL TIME
28 @ 120° + 7°
1933
0800
2733 SAME DAY GREENWICH
(2400)
0333 GMT 9-4-86

FIGURE 7.1 ☆

time by plus or minus one hour when crossing into another time zone. The width of each time zone is 15°, and the time zones are measured 7°30′ in either direction from the middle of each time zone. The middle of each time zone is every 15° of longitude on the surface of the earth.

When sailing west, one hour is lost every 15° of longitude, or at the border of the next time zone. When sailing east one hour is gained every 15° of longitude or at the border of the next time zone. When on the California coast it is eight hours earlier than Greenwich. When in Hawaii it is ten hours earlier than Greenwich and two hours earlier than California. I do not keep daylight savings time aboard as it would have to be converted to local time and would only confuse matters.

It is necessary to become familiar with *full-hour meridians*. A full-hour meridian is any meridian that is a multiple of 15°, east or west of Greenwich. Examples of full hour meridians would be 0°, 15°,

30°, 120°, 135°, 150°, 300°, 330°, etc. The full-hour meridians represent the positions of a full hour of change in time.

Open the current year almanac to the first page of the yellow pages headed: CONVERSION OF ARC TO TIME. The tables are helpful for converting expressions of arc to their equivalent in time and expressions of time to their equivalent in arc. For example, the earth rotates 15° in 1 hour, 1° in 4 minutes or in 8 minutes the earth turns 2° and the earth turns 1′ in 4 seconds.

In the lower right hand corner of your worksheet write SUNSET.

Go back to the almanac and go down the column labeled: Lat. in the lower right hand half of the page to 35° N and 30° N. 33° N is close enough to our D.R. Lat. for interpolating purposes.

An intermediate line of information for 33° N would be ⅗ths of the difference of the times given for 35° N and 30° N.

The times for *civil twilight*, when the sun is from 0°50′ to 6° below the horizon and *nautical twilight*, when the sun is from 0°50′ to 12° below the horizon, are given in local time at the nearest full-hour meridian for the corresponding latitudes.

Go across to the column headed: Twilight Nautical and obtain 1922 and 1915 respectively. Subtract 1915 from 1922; the difference is 7 minutes. Multiply ⅗ × 7 = 4.2 minutes or 4 minutes rounded. Add 4 minutes to 1915, obtaining and recording: 1919 TW NAUT as shown.

Go back one column to the column headed: Twilight Civil, and obtain 1852 and 1846 respectively. Subtract 1846 from 1852; the difference is 6 minutes. Multiply ⅗ × 6 = 3.6 minutes or 4 minutes rounded. Add 4 minutes to 1846, obtaining and recording: 1850 TW CIVIL as shown.

Subtract 1850 from 1919; the remainder of 29 minutes is called the *working period.* The working period is the length of time it is possible to observe the navigational stars, planets, and the horizon at the same time.

Divide the working period by two, rounding to the nearest minute to obtain the middle of the working period. The middle of the working period is the best time to observe Polaris and the horizon.

Add the 15 minutes obtained to 1850 TW CIVIL and record: 1905 LOCAL TIME. The Local Time: 1905 hours is obtained for Lat. 33° N at the nearest full hour meridian: 120° W.

Recall that the DR position was about 16 nautical miles less than the 127° meridian earlier today at 1155. Progress has been slow, but according to the log, we have sailed past the 127° meridian.

Go to the CONVERSION OF ARC TO TIME table in the almanac.

Go down the first column to 7°. The equivalent time for 7° is 28 minutes. Record: 28 @ 120° + 7°.

Add the 28 minutes to 1905 LOCAL TIME, obtaining and recording: 1933. This is the local time @ 127°. Do not be concerned with the few minutes of arc past the 127° meridian. These few minutes of arc amount to seconds during the working period.

Go to the third column in the CONVERSION OF ARC TO TIME table and at the 120° full-hour meridian the conversion corresponds to 8 hours. At the 120° full-hour meridian we are 8 hours from Greenwich and in West Longitude; therefore, record: 0800 and add, obtaining and recording: 2733 SAME DAY GREENWICH.

Because this is the next day at Greenwich, subtract 2400 hours, obtaining and recording: 0333 GMT, 9/4/86.

Advance the assumed position to 1933 hours and plot it on the plotting sheet as shown.

The above calculations were accomplished approximately one hour before the sight of Polaris was taken.

Around 1930 hours we started listening to Stations WWV and WWVH and the stopwatch was started at 3h 31m.

On deck, the sextant was checked for index error. There wasn't any index error. Using the DR latitude as an approximation of our actual latitude I preset the sextant to 33°27′.

Taking magnetic variation into consideration, the binnacle compass provided an approximation of the azimuth of Polaris. As I looked in a northerly direction through the telescope of the sextant, Polaris was sighted just below the horizon. I was able to verify the position of Polaris by observing that it was in alignment with the pouring end of the Big Dipper (Ursa Major).

My wife informed me that there were almost two minutes on the stopwatch.

From my position at the port rail aft I said "mark" when Polaris touched the horizon as seen in the horizon glass of the sextant. We recorded: Stop Watch: 2m 2s and Hs: 33°01′.

Record and add the times in the Time block.

Open the almanac to the inside front cover or white card labelled: ALTITUDE CORRECTION TABLES 10°–90°—SUN, STARS, PLANETS.

My eye was approximately 11 feet above the water. Determine the dip and record it in the usual manner.

Calculate and record the apparent altitude: Ha: 32°57.8′.

Go to the STARS AND PLANETS columns in the middle of the page. Go down the first column headed App. Alt. Ha is between 32°00′ and 33°45′. Record: Corr.: −1.5′. Star and planet corrections in the App. Alt. column are always negative.

The bodies are small and appear as a point light source therefore, unlike the sun and moon, there are no upper or lower limb corrections to be concerned with.

Because the apparent altitude was less than 50°, go to the tables headed: A4 ALTITUDE CORRECTION TABLES—ADDITIONAL CORRECTIONS, ADDITIONAL REFRACTION CORRECTIONS FOR NON-STANDARD CONDITIONS.

The temperature was 70°F and the pressure was 1030 mb.

Go across the top line of the graph to 70° and down to the place where 1030 would intersect. The intersection would be in the H column. Go down the column headed: H, to 30° and 35°. The correction would be 0.0′, i.e., there was no correction for non-standard conditions.

Place an asterisk as shown, in the space: Add'l Corr., because a correction essential to Polaris will be inserted in this space.

Open the almanac to the left hand white page headed: 1986 SEPTEMBER 4,5,6, (THUR., FRI., SAT.) or go to Figure 7.2.

Go to the first two columns on the far left hand side of the page.

The first column is the same as on the right hand page for GMT and requires no further explanation.

The second column headed: ARIES describes a meridian in space for measuring the angular positions of all the stars. For Polaris, we will be concerned with obtaining the LHA of Aries only.

The symbol for Aries is the ram's head. The ram's head is entered on the worksheet next to GHA Hours, Min & Sec., and LHA.

Go down the column for GMT to the 4d, THURSDAY to 3 hours and record: GHA Hours: 27°56.6′.

Go to the 33 m page in the yellow pages. Go down the column to 2s and go across to the column headed: ARIES. Record: Min. & Sec.: 8°16.9′.

Put a slash through the SHA space. Although Polaris is a star, Polaris is not treated like the other stars used for celestial navigation because of its unique position near the polar axis.

The technique for obtaining the LHA of Aries is the same as

174 1986 SEPTEMBER 4, 5, 6 (THURS., FRI., SAT.)

G.M.T. (UT) d h	ARIES G.H.A.	VENUS −4.1 G.H.A.	VENUS Dec.	MARS −1.3 G.H.A.	MARS Dec.	JUPITER −2.4 G.H.A.	JUPITER Dec.	SATURN +0.7 G.H.A.	SATURN Dec.
	° ′	° ′	° ′	° ′	° ′	° ′	° ′	° ′	° ′
4 00	342 49.2	138 52.8	S13 06.0	56 09.5	S27 18.2	352 33.1	S 5 49.5	100 48.3	S19 12.3
01	357 51.7	153 53.2	07.1	71 11.2	18.0	7 35.9	49.7	115 50.7	12.3
02	12 54.1	168 53.6	08.1	86 12.9	17.8	22 38.6	49.8	130 53.0	12.4
03	27 56.6	183 54.0	· · 09.2	101 14.6	· · 17.6	37 41.4	· · 49.9	145 55.4	· · 12.4
04	42 59.1	198 54.4	10.3	116 16.3	17.4	52 44.1	50.1	160 57.7	12.4
05	58 01.5	213 54.9	11.4	131 17.9	17.2	67 46.9	50.2	176 00.1	12.4
06	73 04.0	228 55.3	S13 12.4	146 19.6	S27 17.1	82 49.7	S 5 50.3	191 02.4	S19 12.5
07	88 06.5	243 55.7	13.5	161 21.3	16.9	97 52.4	50.5	206 04.8	12.5
T 08	103 08.9	258 56.1	14.6	176 23.0	16.7	112 55.2	50.6	221 07.1	12.5
H 09	118 11.4	273 56.5	· · 15.7	191 24.6	· · 16.5	127 58.0	· · 50.7	236 09.5	· · 12.6
U 10	133 13.8	288 56.9	16.8	206 26.3	16.3	143 00.7	50.9	251 11.8	12.6
R 11	148 16.3	303 57.4	17.8	221 28.0	16.1	158 03.5	51.0	266 14.2	12.6
S 12	163 18.8	318 57.8	S13 18.9	236 29.6	S27 15.9	173 06.3	S 5 51.1	281 16.5	S19 12.7
D 13	178 21.2	333 58.2	20.0	251 31.3	15.7	188 09.0	51.3	296 18.9	12.7
A 14	193 23.7	348 58.6	21.0	266 33.0	15.5	203 11.8	51.4	311 21.2	12.7
Y 15	208 26.2	3 59.1	· · 22.1	281 34.6	· · 15.3	218 14.6	· · 51.5	326 23.6	· · 12.7
16	223 28.6	18 59.5	23.2	296 36.3	15.1	233 17.3	51.6	341 25.9	12.8
17	238 31.1	33 59.9	24.3	311 38.0	14.9	248 20.1	51.8	356 28.3	12.8
18	253 33.6	49 00.3	S13 25.3	326 39.6	S27 14.7	263 22.9	S 5 51.9	11 30.6	S19 12.8
19	268 36.0	64 00.8	26.4	341 41.3	14.5	278 25.6	52.0	26 33.0	12.9
20	283 38.5	79 01.2	27.5	356 42.9	14.3	293 28.4	52.2	41 35.3	12.9
21	298 40.9	94 01.6	· · 28.6	11 44.6	· · 14.1	308 31.1	· · 52.3	56 37.7	· · 12.9
22	313 43.4	109 02.0	29.6	26 46.3	13.9	323 33.9	52.4	71 40.0	13.0
23	328 45.9	124 02.5	30.7	41 47.9	13.7	338 36.7	52.6	86 42.4	13.0
5 00	343 48.3	139 02.9	S13 31.8	56 49.6	S27 13.5	353 39.4	S 5 52.7	101 44.7	S19 13.0
01	358 50.8	154 03.3	32.8	71 51.2	13.3	8 42.2	52.8	116 47.1	13.0
02	13 53.3	169 03.8	33.9	86 52.9	13.1	23 45.0	53.0	131 49.4	13.1
03	28 55.7	184 04.2	· · 35.0	101 54.5	· · 12.9	38 47.7	· · 53.1	146 51.7	· · 13.1
04	43 58.2	199 04.6	36.0	116 56.2	12.7	53 50.5	53.2	161 54.1	13.1
05	59 00.7	214 05.0	37.1	131 57.8	12.5	68 53.3	53.4	176 56.4	13.2
06	74 03.1	229 05.5	S13 38.2	146 59.5	S27 12.3	83 56.0	S 5 53.5	191 58.8	S19 13.2
07	89 05.6	244 05.9	39.2	162 01.1	12.1	98 58.8	53.6	207 01.1	13.2
08	104 08.1	259 06.3	40.3	177 02.8	11.9	114 01.6	53.8	222 03.5	13.3
F 09	119 10.5	274 06.8	· · 41.4	192 04.4	· · 11.7	129 04.3	· · 53.9	237 05.8	· · 13.3
R 10	134 13.0	289 07.2	42.4	207 06.0	11.5	144 07.1	54.0	252 08.2	13.3
I 11	149 15.4	304 07.7	43.5	222 07.7	11.3	159 09.9	54.2	267 10.5	13.4
D 12	164 17.9	319 08.1	S13 44.5	237 09.3	S27 11.1	174 12.6	S 5 54.3	282 12.9	S19 13.4
A 13	179 20.4	334 08.5	45.6	252 11.0	10.9	189 15.4	54.4	297 15.2	13.4
Y 14	194 22.8	349 09.0	46.7	267 12.6	10.7	204 18.2	54.6	312 17.6	13.4
15	209 25.3	4 09.4	· · 47.7	282 14.2	· · 10.5	219 20.9	· · 54.7	327 19.9	· · 13.5
16	224 27.8	19 09.9	48.8	297 15.9	10.3	234 23.7	54.8	342 22.2	13.5
17	239 30.2	34 10.3	49.8	312 17.5	10.1	249 26.5	55.0	357 24.6	13.5
18	254 32.7	49 10.7	S13 50.9	327 19.1	S27 09.9	264 29.2	S 5 55.1	12 26.9	S19 13.6
19	269 35.2	64 11.2	52.0	342 20.8	09.7	279 32.0	55.2	27 29.3	13.6
20	284 37.6	79 11.6	53.0	357 22.4	09.5	294 34.8	55.3	42 31.6	13.6
21	299 40.1	94 12.1	· · 54.1	12 24.0	· · 09.3	309 37.5	· · 55.5	57 34.0	· · 13.7
22	314 42.6	109 12.5	55.1	27 25.7	09.1	324 40.3	55.6	72 36.3	13.7
23	329 45.0	124 12.9	56.2	42 27.3	08.9	339 43.1	55.7	87 38.7	13.7
6 00	344 47.5	139 13.4	S13 57.3	57 28.9	S27 08.7	354 45.8	S 5 55.9	102 41.0	S19 13.8
01	359 49.9	154 13.8	58.3	72 30.5	08.5	9 48.6	56.0	117 43.3	13.8
02	14 52.4	169 14.3	13 59.4	87 32.2	08.3	24 51.4	56.1	132 45.7	13.8
03	29 54.9	184 14.7	14 00.4	102 33.8	· · 08.1	39 54.1	· · 56.3	147 48.0	· · 13.8
04	44 57.3	199 15.2	01.5	117 35.4	07.9	54 56.9	56.4	162 50.4	13.9
05	59 59.8	214 15.6	02.5	132 37.0	07.7	69 59.7	56.5	177 52.7	13.9
06	75 02.3	229 16.1	S14 03.6	147 38.6	S27 07.4	85 02.4	S 5 56.7	192 55.1	S19 13.9
07	90 04.7	244 16.5	04.6	162 40.3	07.2	100 05.2	56.8	207 57.4	14.0
S 08	105 07.2	259 17.0	05.7	177 41.9	07.0	115 08.0	56.9	222 59.7	14.0
A 09	120 09.7	274 17.4	· · 06.7	192 43.5	· · 06.8	130 10.7	· · 57.1	238 02.1	· · 14.0
T 10	135 12.1	289 17.9	07.8	207 45.1	06.6	145 13.5	57.2	253 04.4	14.1
U 11	150 14.6	304 18.3	08.8	222 46.7	06.4	160 16.3	57.3	268 06.8	14.1
R 12	165 17.0	319 18.8	S14 09.9	237 48.3	S27 06.2	175 19.0	S 5 57.5	283 09.1	S19 14.1
D 13	180 19.5	334 19.3	10.9	252 49.9	06.0	190 21.8	57.6	298 11.5	14.2
A 14	195 22.0	349 19.7	12.0	267 51.5	05.8	205 24.6	57.7	313 13.8	14.2
Y 15	210 24.4	4 20.2	· · 13.0	282 53.2	· · 05.6	220 27.4	· · 57.9	328 16.1	· · 14.2
16	225 26.9	19 20.6	14.1	297 54.8	05.4	235 30.1	58.0	343 18.5	14.3
17	240 29.4	34 21.1	15.1	312 56.4	05.2	250 32.9	58.1	358 20.8	14.3
18	255 31.8	49 21.5	S14 16.2	327 58.0	S27 05.0	265 35.7	S 5 58.3	13 23.2	S19 14.3
19	270 34.3	64 22.0	17.2	342 59.6	04.8	280 38.4	58.4	28 25.5	14.3
20	285 36.8	79 22.5	18.3	358 01.2	04.6	295 41.2	58.5	43 27.8	14.4
21	300 39.2	94 22.9	· · 19.3	13 02.8	· · 04.3	310 44.0	· · 58.7	58 30.2	· · 14.4
22	315 41.7	109 23.4	20.4	28 04.4	04.1	325 46.7	58.8	73 32.5	14.4
23	330 44.2	124 23.8	21.4	43 06.0	03.9	340 49.5	58.9	88 34.9	14.5
Mer. Pass. (h m)	1 04.6	v 0.4	d 1.1	v 1.6	d 0.2	v 2.8	d 0.1	v 2.3	d 0.0

STARS Name	S.H.A.	Dec.
	° ′	° ′
Acamar	315 33.5	S40 21.1
Achernar	335 41.1	S57 18.0
Acrux	173 33.2	S63 01.6
Adhara	255 28.8	S28 56.8
Aldebaran	291 12.8	N16 29.2
Alioth	166 38.7	N56 02.1
Alkaid	153 15.0	N49 23.0
Al Na'ir	28 08.6	S47 01.6
Alnilam	276 07.1	S 1 12.3
Alphard	218 16.4	S 8 35.8
Alphecca	126 28.3	N26 45.7
Alpheratz	358 04.4	N29 01.0
Altair	62 27.9	N 8 50.0
Ankaa	353 35.2	S42 22.5
Antares	112 51.3	S26 24.3
Arcturus	146 14.5	N19 15.2
Atria	108 11.6	S69 00.7
Avior	234 27.1	S59 27.6
Bellatrix	278 53.9	N 6 20.5
Betelgeuse	271 23.4	N 7 24.6
Canopus	264 05.4	S52 40.8
Capella	281 04.7	N45 59.1
Deneb	49 45.1	N45 14.0
Denebola	182 54.7	N14 39.0
Diphda	349 15.9	S18 03.4
Dubhe	194 16.8	N61 49.5
Elnath	278 38.5	N28 35.9
Eltanin	90 55.5	N51 29.6
Enif	34 06.8	N 9 48.8
Fomalhaut	15 45.9	S29 41.5
Gacrux	172 24.5	S57 02.3
Gienah	176 13.7	S17 28.0
Hadar	149 17.6	S60 18.7
Hamal	328 23.6	N23 24.1
Kaus Aust.	84 10.6	S34 23.7
Kochab	137 19.2	N74 12.8
Markab	13 58.4	N15 08.0
Menkar	314 36.2	N 4 02.5
Menkent	148 32.0	S36 18.4
Miaplacidus	221 45.3	S69 39.4
Mirfak	309 09.6	N49 48.8
Nunki	76 23.3	S26 19.0
Peacock	53 50.6	S56 46.9
Pollux	243 52.8	N28 03.7
Procyon	245 21.2	N 5 15.8
Rasalhague	96 25.3	N12 34.2
Regulus	208 05.4	N12 02.2
Rigel	281 31.7	S 8 12.7
Rigil Kent.	140 20.3	S60 47.0
Sabik	102 35.9	S15 42.6
Schedar	350 03.7	N56 27.8
Shaula	96 49.5	S37 05.9
Sirius	258 51.9	S16 41.5
Spica	158 53.0	S11 05.5
Suhail	223 08.0	S43 22.4
Vega	80 52.6	N38 46.4
Zuben'ubi	137 28.2	S15 59.2

	S.H.A.	Mer. Pass.
	° ′	h m
Venus	155 14.6	14 43
Mars	73 01.2	20 10
Jupiter	9 51.1	0 25
Saturn	117 56.4	17 10

FIGURE 7.2 ☆ *NAUTICAL ALMANAC*, LEFT HAND DAILY PAGE (TYPICAL)

in previous examples. Calculate the LHA of Aries. I calculated and recorded: LHA of Aries: 269°.

Open the almanac to the last few pages in the white page section. In these last pages you will find the tables: POLARIS (POLE STAR) TABLES, 1986 FOR DETERMINING LATITUDE FROM SEXTANT ALTITUDE AND FOR AZIMUTH or go to Figure 7.3.

The purpose of the Polaris tables is to correct for the appearance of a wobbling effect of the star around the polar axis. That is, a correction must be applied to compensate for the fact that Polaris is not precisely north of the North Pole and in effect rotates in a small, irregular path around the extension of the polar axis.

The formula for obtaining the correction is:

$$\text{LAT} = \text{APP/ALT} - 1° + a_0 + a_1 + a_2.$$

The Polaris table gives a series of corrections Headed: LHA of Aries, Lat., Month, and Azimuth.

Go across the column headed: L.H.A./ARIES to the column 260°–269°.

Go down this column and record the three values of *a* used in the formula. The first group for LHA 269° is: $a_0 = 1°26.6'$; the second group for Lat. using 30° is: $a_1 = 0.5'$; and the third group for Month using Sept. is: $a_2 = 0.9'$.

Add the three figures and subtract 1° to obtain and record: 28′ in the space labelled: Add'l Corr.* as shown.

Adding the corrections algebraically to Ha and rounding off to the nearest minute, we obtain and record: Ho: 33°24′, the latitude.

Plot the LOP horizontally and label it: 1933/POLARIS.

Label the resulting position horizontally: 1933 R FIX. We then transferred our latitude and longitude to the "big chart".

There is a faster and easier way to obtain the correction for Polaris. Go to Figure 7.4 headed: TABLE 6—CORRECTION (Q) FOR POLARIS. This insert is from *Vol. 1., Pub. No. 249 Sight Reduction Tables for Air Navigation for Epoch 1985*.

Table 6 is organized so that for any LHA of Aries there is a corresponding Q correction. The LHA of Aries calculated was 269° and between LHA values 267°57′ and 269°26′ in Table 6 find the correction: +28′. The Table 6 correction was the same as the almanac method; any difference will be of no significance to practical small-craft navigation.

I felt it was important to teach the method using the almanac

POLARIS (POLE STAR) TABLES, 1986

FOR DETERMINING LATITUDE FROM SEXTANT ALTITUDE AND FOR AZIMUTH

L.H.A. ARIES	240°–249°	250°–259°	260°–269°	270°–279°	280°–289°	290°–299°	300°–309°	310°–319°	320°–329°	330°–339°	340°–349°	350°–359°
	a_0	a_0	a_0	a_0	a_0	a_0	a_0	a_0	a_0	a_0	a_0	a_0
°	° ′	° ′	° ′	° ′	° ′	° ′	° ′	° ′	° ′	° ′	° ′	° ′
0	1 41·9	1 37·7	1 32·3	1 25·9	1 18·8	1 11·0	1 02·8	0 54·5	0 46·3	0 38·4	0 31·2	0 24·9
1	41·5	37·2	31·7	25·3	18·0	10·1	01·9	53·6	45·5	37·7	30·5	24·3
2	41·1	36·7	31·1	24·6	17·2	09·3	01·1	52·8	44·7	36·9	29·9	23·7
3	40·7	36·2	30·5	23·9	16·5	08·5	1 00·3	52·0	43·9	36·2	29·2	23·1
4	40·3	35·7	29·9	23·2	15·7	07·7	0 59·4	51·1	43·1	35·5	28·6	22·6
5	1 39·9	1 35·1	1 29·2	1 22·4	1 14·9	1 06·9	0 58·6	0 50·3	0 42·3	0 34·7	0 27·9	0 22·0
6	39·5	34·6	28·6	21·7	14·1	06·1	57·8	49·5	41·5	34·0	27·3	21·5
7	39·1	34·0	28·0	21·0	13·3	05·2	56·9	48·7	40·7	33·3	26·7	21·0
8	38·6	33·5	27·3	20·3	12·6	04·4	56·1	47·9	40·0	32·6	26·1	20·5
9	38·1	32·9	26·6	19·5	11·8	03·6	55·3	47·1	39·2	31·9	25·4	20·0
10	1 37·7	1 32·3	1 25·9	1 18·8	1 11·0	1 02·8	0 54·5	0 46·3	0 38·4	0 31·2	0 24·9	0 19·5
Lat.	a_1	a_1	a_1	a_1	a_1	a_1	a_1	a_1	a_1	a_1	a_1	a_1
°	′	′	′	′	′	′	′	′	′	′	′	′
0	0·5	0·4	0·4	0·3	0·2	0·2	0·2	0·2	0·3	0·3	0·4	0·4
10	·5	·5	·4	·3	·3	·3	·3	·3	·3	·4	·4	·5
20	·5	·5	·4	·4	·4	·3	·3	·3	·4	·4	·4	·5
30	·5	·5	·5	·4	·4	·4	·4	·4	·4	·4	·5	·5
40	0·6	0·6	0·5	0·5	0·5	0·5	0·5	0·5	0·5	0·5	0·5	0·6
45	·6	·6	·6	·6	·5	·5	·5	·5	·5	·6	·6	·6
50	·6	·6	·6	·6	·6	·6	·6	·6	·6	·6	·6	·6
55	·6	·6	·6	·7	·7	·7	·7	·7	·7	·7	·6	·6
60	·6	·7	·7	·7	·8	·8	·8	·8	·8	·7	·7	·7
62	0·7	0·7	0·7	0·8	0·8	0·8	0·8	0·8	0·8	0·8	0·7	0·7
64	·7	·7	·8	·8	·9	·9	·9	·9	·8	·8	·8	·7
66	·7	·7	·8	·9	0·9	0·9	0·9	0·9	0·9	·9	·8	·7
68	0·7	0·8	0·9	0·9	1·0	1·0	1·0	1·0	1·0	0·9	0·8	0·8
Month	a_2	a_2	a_2	a_2	a_2	a_2	a_2	a_2	a_2	a_2	a_2	a_2
	′	′	′	′	′	′	′	′	′	′	′	′
Jan.	0·5	0·5	0·5	0·5	0·5	0·5	0·6	0·6	0·6	0·6	0·7	0·7
Feb.	·4	·4	·4	·4	·4	·4	·4	·4	·5	·5	·5	·6
Mar.	·4	·4	·3	·3	·3	·3	·3	·3	·3	·4	·4	·4
Apr.	0·5	0·4	0·4	0·3	0·3	0·3	0·2	0·2	0·2	0·2	0·3	0·3
May	·6	·6	·5	·4	·4	·3	·3	·2	·2	·2	·2	·2
June	·8	·7	·6	·6	·5	·4	·4	·3	·3	·2	·2	·2
July	0·9	0·8	0·8	0·7	0·7	0·6	0·5	0·5	0·4	0·4	0·3	0·3
Aug.	·9	·9	·9	·8	·8	·8	·7	·6	·6	·5	·5	·4
Sept.	·9	·9	·9	·9	·9	·9	·8	·8	·8	·7	·7	·6
Oct.	0·8	0·8	0·9	0·9	0·9	0·9	0·9	0·9	0·9	0·9	0·8	0·8
Nov.	·6	·7	·8	·8	·9	·9	1·0	1·0	1·0	1·0	1·0	1·0
Dec.	0·5	0·5	0·6	0·7	0·8	0·8	0·9	1·0	1·0	1·0	1·0	1·1
Lat.	AZIMUTH											
°	°	°	°	°	°	°	°	°	°	°	°	°
0	0·4	0·5	0·6	0·7	0·8	0·8	0·8	0·8	0·7	0·7	0·6	0·5
20	0·4	0·5	0·7	0·7	0·8	0·8	0·8	0·8	0·8	0·7	0·6	0·5
40	0·5	0·7	0·8	0·9	1·0	1·0	1·0	1·0	1·0	0·9	0·8	0·7
50	0·6	0·8	0·9	1·1	1·2	1·2	1·2	1·2	1·2	1·1	0·9	0·8
55	0·7	0·9	1·1	1·2	1·3	1·4	1·4	1·4	1·3	1·2	1·1	0·9
60	0·8	1·0	1·2	1·4	1·5	1·6	1·6	1·6	1·5	1·4	1·2	1·0
65	0·9	1·2	1·4	1·6	1·8	1·8	1·9	1·9	1·8	1·6	1·5	1·2

$$\text{Latitude} = \text{Apparent altitude (corrected for refraction)} - 1^\circ + a_0 + a_1 + a_2$$

The table is entered with L.H.A. Aries to determine the column to be used; each column refers to a range of 10°. a_0 is taken, with mental interpolation, from the upper table with the units of L.H.A. Aries in degrees as argument; a_1, a_2 are taken, without interpolation, from the second and third tables with arguments latitude and month respectively. a_0, a_1, a_2 are always positive. The final table gives the azimuth of *Polaris*.

FIGURE 7.3 ☆ *NAUTICAL ALMANAC*, POLARIS TABLES (TYPICAL)

TABLE 6—CORRECTION (*Q*) FOR *POLARIS*

LHA ♈	*Q*	LHA ♈	*Q*	LHA ♈	*Q*	LHA ♈	*Q*	LHA ♈	*Q*	LHA ♈	*Q*	LHA ♈	*Q*	LHA ♈	*Q*
° ′	′	° ′	′	° ′	′	° ′	′	° ′	′	° ′	′	° ′	′	° ′	′
359 40	−40	84 18	−30	118 16	− 4	150 19	+22	205 44	+48	276 27	+22	308 39	− 4	342 18	−30
1 49	−41	85 49	−29	119 28	− 3	151 40	+23	222 23	+47	277 48	+21	309 51	− 5	343 49	−31
4 06	−42	87 18	−28	120 40	− 2	153 02	+24	228 31	+46	279 08	+20	311 03	− 6	345 23	−32
6 33	−43	88 46	−27	121 52	− 1	154 25	+25	232 45	+45	280 27	+19	312 15	− 7	346 59	−33
9 14	−44	90 13	−26	123 03	0	155 49	+26	236 13	+44	281 45	+18	313 27	− 8	348 37	−34
12 12	−45	91 37	−25	124 16	+ 1	157 14	+27	239 14	+43	283 03	+17	314 40	− 9	350 18	−35
15 37	−46	93 01	−24	125 27	+ 2	158 41	+28	241 56	+42	284 19	+16	315 52	−10	352 02	−36
19 48	−47	94 23	−23	126 39	+ 3	160 10	+29	244 25	+41	285 35	+15	317 06	−11	353 50	−37
25 51	−48	95 45	−22	127 51	+ 4	161 40	+30	246 44	+40	286 51	+14	318 19	−12	355 41	−38
42 16	−47	97 05	−21	129 03	+ 5	163 12	+31	248 54	+39	288 06	+13	319 33	−13	357 38	−39
48 19	−46	98 24	−20	130 15	+ 6	164 47	+32	250 58	+38	289 21	+12	320 47	−14	359 40	−40
52 30	−45	99 43	−19	131 27	+ 7	166 23	+33	252 56	+37	290 35	+11	322 02	−15	1 49	−41
55 55	−44	101 01	−18	132 40	+ 8	168 02	+34	254 49	+36	291 48	+10	323 17	−16	4 06	−42
58 53	−43	102 18	−17	133 52	+ 9	169 44	+35	256 38	+35	293 02	+ 9	324 33	−17	6 33	−43
61 34	−42	103 34	−16	135 05	+10	171 29	+36	258 23	+34	294 15	+ 8	325 49	−18	9 14	−44
64 01	−41	104 50	−15	136 19	+11	173 18	+37	260 05	+33	295 27	+ 7	327 06	−19	12 12	−45
66 18	−40	106 05	−14	137 32	+12	175 11	+38	261 44	+32	296 40	+ 6	328 24	−20	15 37	−46
68 27	−39	107 20	−13	138 46	+13	177 09	+39	263 20	+31	297 52	+ 5	329 43	−21	19 48	−47
70 29	−38	108 34	−12	140 01	+14	179 13	+40	264 55	+30	299 04	+ 4	331 02	−22	25 51	−48
72 26	−37	109 48	−11	141 16	+15	181 23	+41	266 27	+29	300 16	+ 3	332 22	−23	42 16	−47
74 17	−36	111 01	−10	142 32	+16	183 42	+42	267 57	+28	301 28	+ 2	333 44	−24	48 19	−46
76 05	−35	112 15	− 9	143 48	+17	186 11	+43	269 26	+27	302 40	+ 1	335 06	−25	52 30	−45
77 49	−34	113 27	− 8	145 04	+18	188 53	+44	270 53	+26	303 51	0	336 30	−26	55 55	−44
79 30	−33	114 40	− 7	146 22	+19	191 54	+45	272 18	+25	305 04	− 1	337 54	−27	58 53	−43
81 08	−32	115 52	− 6	147 40	+20	195 22	+46	273 42	+24	306 15	− 2	339 21	−28	61 34	−42
82 44	−31	117 04	− 5	148 59	+21	199 36	+47	275 05	+23	307 27	− 3	340 49	−29	64 01	−41
84 18		118 16		150 19		205 44		276 27		308 39		342 18		66 18	

The above table, which does *not* include refraction, gives the quantity *Q* to be applied to the corrected sextant altitude of *Polaris* to give the latitude of the observer. In critical cases ascend.

Polaris: Mag. 2·1, SHA 325° 56′, Dec. N 89° 12′·0

TABLE 7—AZIMUTH OF *POLARIS*

LHA ♈	Latitude 0°	30°	50°	55°	60°	65°	70°	LHA ♈	Latitude 0°	30°	50°	55°	60°	65°	70°
°	°	°	°	°	°	°	°	°	°	°	°	°	°	°	°
0	0·4	0·5	0·7	0·8	0·9	1·1	1·4	180	359·6	359·5	359·3	359·2	359·1	359·0	358·7
10	0·3	0·4	0·5	0·6	0·7	0·8	1·0	190	359·7	359·6	359·5	359·4	359·4	359·2	359·1
20	0·2	0·2	0·3	0·3	0·4	0·5	0·6	200	359·8	359·8	359·7	359·7	359·6	359·6	359·5
30	0·1	0·1	0·1	0·1	0·1	0·1	0·2	210	359·9	359·9	359·9	359·9	359·9	359·9	359·8
40	359·9	359·9	359·9	359·9	359·8	359·8	359·7	220	0·1	0·1	0·1	0·1	0·2	0·2	0·2
50	359·8	359·7	359·7	359·6	359·6	359·5	359·3	230	0·2	0·3	0·3	0·4	0·4	0·5	0·6
60	359·7	359·6	359·4	359·4	359·3	359·1	358·9	240	0·3	0·4	0·5	0·6	0·7	0·8	1·0
70	359·5	359·5	359·3	359·2	359·0	358·9	358·6	250	0·5	0·5	0·7	0·8	0·9	1·1	1·3
80	359·4	359·3	359·1	359·0	358·8	358·6	358·3	260	0·6	0·7	0·9	1·0	1·1	1·3	1·6
90	359·3	359·2	359·0	358·8	358·7	358·4	358·0	270	0·7	0·8	1·0	1·1	1·3	1·5	1·9
100	359·3	359·2	358·9	358·7	358·5	358·3	357·8	280	0·7	0·8	1·1	1·3	1·4	1·7	2·1
110	359·2	359·1	358·8	358·6	358·4	358·2	357·7	290	0·8	0·9	1·2	1·3	1·5	1·8	2·2
120	359·2	359·1	358·8	358·6	358·4	358·1	357·7	300	0·8	0·9	1·2	1·4	1·6	1·9	2·3
130	359·2	359·1	358·8	358·6	358·4	358·1	357·7	310	0·8	0·9	1·2	1·4	1·6	1·9	2·3
140	359·2	359·1	358·8	358·7	358·5	358·2	357·8	320	0·8	0·9	1·2	1·3	1·5	1·8	2·3
150	359·3	359·2	358·9	358·8	358·6	358·3	357·9	330	0·7	0·8	1·1	1·3	1·5	1·7	2·1
160	359·4	359·3	359·0	358·9	358·7	358·5	358·1	340	0·6	0·8	1·0	1·1	1·3	1·6	1·9
170	359·4	359·4	359·1	359·0	358·9	358·7	358·4	350	0·6	0·6	0·9	1·0	1·1	1·3	1·7
180	359·6	359·5	359·3	359·2	359·1	359·0	358·7	360	0·4	0·5	0·7	0·8	0·9	1·1	1·4

When Cassiopeia is left (right), *Polaris* is west (east).

FIGURE 7.4 ☆ *VOL. I, PUB. NO. 249*, CORRECTION TABLE FOR POLARIS (TYPICAL)

even though the Q factor method was much faster and simpler. The almanac could be all that is available to you in a survival situation.

The new course is 249° true. Draw the DR course line from the middle of the fix and label it C 249.

STAR SELECTION, NORTH PACIFIC

EIGHT

THE STAR FIX I

We are well into our fourth day at sea since departing Morro Bay. The winds are moderate, the sea has settled into a fairly comfortable, regular pattern of waves and swells, and the temperature has been rising. I am always amazed how pleasant conditions can be after the coastal influences of fog, cold weather, and confused seas have been left astern.

About two hundred yards from us, off the port side, we sighted a sperm whale lolling around on the surface. He would blow now and then, and I saw him look our way. We thouroughly enjoy the visits of whales and dolphins at sea. They are good company when they follow us or swim alongside.

We were comfortably exhausted from the activity of the previous day and made a decision to postpone celestial navigation until later in the day. About an hour before sunset we prepared to sight some stars. As evening approached, it appeared that we would have clear skies. I determined the working period and prepared a list of stars to sight.

Begin with an LOP worksheet as shown in Figure 8.1.

Record as follows: No.: 8; Area: NORTH PACIFIC and Date: 9/4/86.

Open the almanac to the right hand white page headed: 1986 SEPTEMBER 4, 5, 6 (THURS., FRI., SAT.) or Figure 8.2.

Write SUNSET in the space provided for calculations on the worksheet. You will be using the tables headed: Twilight Civil and Naut. in the lower right hand portion of the page.

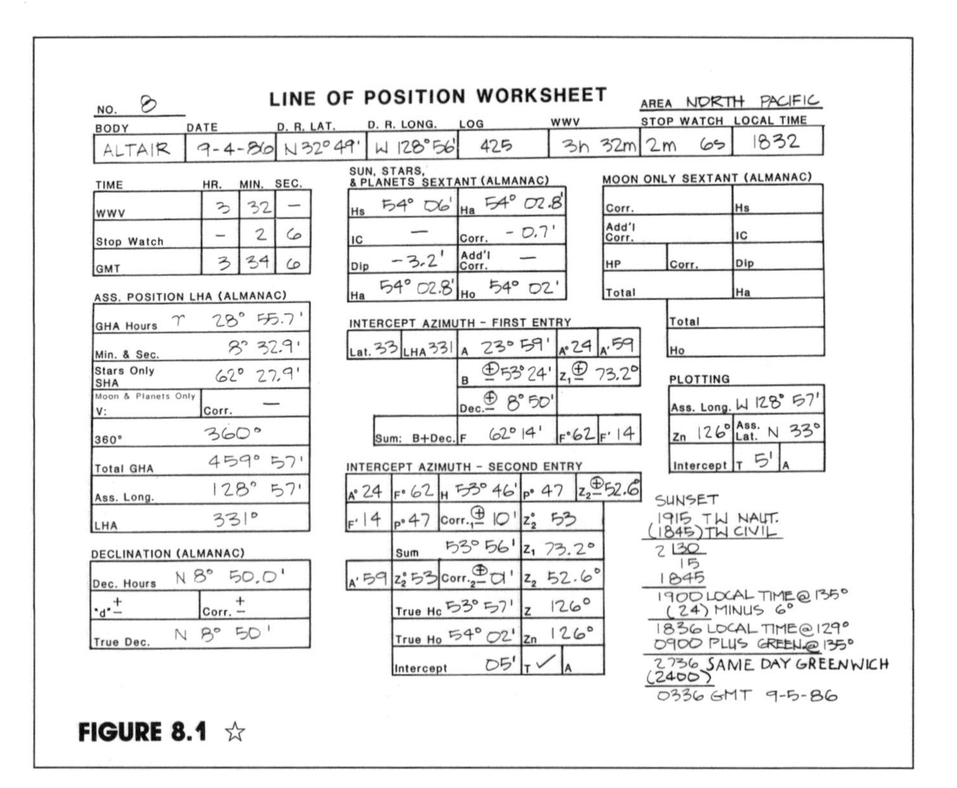

LINE OF POSITION WORKSHEET

NO. 8 AREA NORTH PACIFIC

BODY	DATE	D. R. LAT.	D. R. LONG.	LOG	WWV	STOP WATCH	LOCAL TIME
ALTAIR	9-4-86	N 32° 49'	W 128° 56'	425	3h 32m	2m 6s	1832

TIME	HR.	MIN.	SEC.
WWV	3	32	—
Stop Watch	—	2	6
GMT	3	34	6

ASS. POSITION LHA (ALMANAC)

GHA Hours ♈	28° 55.7'
Min. & Sec.	8° 32.9'
Stars Only SHA	62° 27.9'
Moon & Planets Only V: / Corr.	—
360°	360°
Total GHA	459° 57'
Ass. Long.	128° 57'
LHA	331°

DECLINATION (ALMANAC)

Dec. Hours	N 8° 50.0'
"d" + − / Corr. + −	
True Dec.	N 8° 50'

SUN, STARS, & PLANETS SEXTANT (ALMANAC)

Hs	54° 06'	Ha	54° 02.8'
IC	—	Corr.	− 0.7'
Dip	− 3.2'	Add'l Corr.	—
Ha	54° 02.8'	Ho	54° 02'

INTERCEPT AZIMUTH – FIRST ENTRY

Lat. 33	LHA 331	A 23° 59'	A° 24	A' 59
		B ⊕ 53° 24'	Z_1 ⊕ 73.2°	
		Dec. ⊕ 8° 50'		
Sum: B+Dec.		F 62° 14'	F° 62	F' 14

INTERCEPT AZIMUTH – SECOND ENTRY

A° 24	F° 62	H 53° 46'	P° 47	Z_2 ⊕ 52.6
F' 14	P° 47	$Corr._1$ ⊕ 10'	Z_2° 53	
		Sum 53° 56'	Z_1 73.2°	
A' 59	Z_2° 53	$Corr._2$ ⊕ 01'	Z_2 52.6°	
		True Hc 53° 57'	Z 126°	
		True Ho 54° 02'	Zn 126°	
		Intercept 05'	T ✓	A

MOON ONLY SEXTANT (ALMANAC)

Corr.		Hs
Add'l Corr.		IC
HP	Corr.	Dip
Total		Ha
		Total
		Ho

PLOTTING

Ass. Long.	W 128° 57'
Zn 126°	Ass. Lat. N 33°
Intercept	T 5' A

SUNSET
1915 TW NAUT.
(1845) TW CIVIL
2 130
15
1845
1900 LOCAL TIME @ 135°
(24) MINUS 6°
1836 LOCAL TIME @ 129°
0900 PLUS GREEN. @ 135°
2736 SAME DAY GREENWICH
(2400)
0336 GMT 9-5-86

FIGURE 8.1 ☆

1986 SEPTEMBER 4, 5, 6 (THURS., FRI., SAT.) 175

	G.M.T. (UT)	SUN G.H.A.	SUN Dec.	MOON G.H.A.	MOON v	MOON Dec.	MOON d	MOON H.P.
	d h	° ′	° ′	° ′	′	° ′	′	′
	4 00	180 11.3	N 7 22.5	182 13.6	13.4	N12 03.6	14.3	56.8
	01	195 11.5	21.6	196 46.0	13.3	11 49.3	14.3	56.8
	02	210 11.7	20.7	211 18.3	13.4	11 35.0	14.4	56.8
	03	225 11.9	.. 19.8	225 50.7	13.5	11 20.6	14.5	56.8
	04	240 12.1	18.8	240 23.2	13.4	11 06.1	14.6	56.9
	05	255 12.3	17.9	254 55.6	13.5	10 51.5	14.5	56.9
	06	270 12.5	N 7 17.0	269 28.1	13.5	N10 37.0	14.7	56.9
	07	285 12.7	16.1	284 00.6	13.5	10 22.3	14.7	56.9
T	08	300 12.9	15.2	298 33.1	13.5	10 07.6	14.8	57.0
H	09	315 13.1	.. 14.2	313 05.6	13.6	9 52.8	14.8	57.0
U	10	330 13.3	13.3	327 38.2	13.5	9 38.0	14.9	57.0
R	11	345 13.6	12.4	342 10.7	13.6	9 23.1	14.9	57.0
S	12	0 13.8	N 7 11.5	356 43.3	13.6	N 9 08.2	15.0	57.0
D	13	15 14.0	10.5	11 15.9	13.6	8 53.2	15.1	57.1
A	14	30 14.2	09.6	25 48.5	13.7	8 38.1	15.1	57.1
Y	15	45 14.4	.. 08.7	40 21.2	13.6	8 23.0	15.1	57.1
	16	60 14.6	07.8	54 53.8	13.7	8 07.9	15.2	57.1
	17	75 14.8	06.9	69 26.5	13.7	7 52.7	15.3	57.2
	18	90 15.0	N 7 05.9	83 59.2	13.7	N 7 37.4	15.2	57.2
	19	105 15.2	05.0	98 31.9	13.7	7 22.2	15.4	57.2
	20	120 15.4	04.1	113 04.6	13.7	7 06.8	15.3	57.2
	21	135 15.6	.. 03.2	127 37.3	13.7	6 51.5	15.5	57.3
	22	150 15.8	02.2	142 10.0	13.7	6 36.0	15.4	57.3
	23	165 16.0	01.3	156 42.7	13.8	6 20.6	15.5	57.3
	5 00	180 16.2	N 7 00.4	171 15.5	13.7	N 6 05.1	15.6	57.3
	01	195 16.4	6 59.5	185 48.2	13.8	5 49.5	15.6	57.4
	02	210 16.6	58.5	200 21.0	13.7	5 33.9	15.6	57.4
	03	225 16.8	.. 57.6	214 53.7	13.8	5 18.3	15.6	57.4
	04	240 17.1	56.7	229 26.5	13.7	5 02.7	15.7	57.4
	05	255 17.3	55.8	243 59.2	13.8	4 47.0	15.7	57.4
	06	270 17.5	N 6 54.8	258 32.0	13.7	N 4 31.3	15.8	57.5
	07	285 17.7	53.9	273 04.7	13.8	4 15.5	15.8	57.5
	08	300 17.9	53.0	287 37.5	13.7	3 59.7	15.8	57.5
F	09	315 18.1	.. 52.0	302 10.2	13.8	3 43.9	15.8	57.5
R	10	330 18.3	51.1	316 43.0	13.7	3 28.1	15.9	57.6
I	11	345 18.5	50.2	331 15.7	13.8	3 12.2	15.9	57.6
D	12	0 18.7	N 6 49.3	345 48.5	13.7	N 2 56.3	15.9	57.6
A	13	15 18.9	48.3	0 21.2	13.7	2 40.4	15.9	57.6
Y	14	30 19.1	47.4	14 53.9	13.7	2 24.5	16.0	57.6
	15	45 19.3	.. 46.5	29 26.6	13.7	2 08.5	16.0	57.7
	16	60 19.5	45.6	43 59.3	13.7	1 52.5	16.0	57.7
	17	75 19.8	44.6	58 32.0	13.7	1 36.5	16.0	57.7
	18	90 20.0	N 6 43.7	73 04.7	13.7	N 1 20.5	16.1	57.7
	19	105 20.2	42.8	87 37.4	13.6	1 04.4	16.0	57.7
	20	120 20.4	41.8	102 10.0	13.6	0 48.4	16.1	57.8
	21	135 20.6	.. 40.9	116 42.6	13.7	0 32.3	16.1	57.8
	22	150 20.8	40.0	131 15.3	13.6	0 16.2	16.1	57.8
	23	165 21.0	39.1	145 47.9	13.5	N 0 00.1	16.1	57.8
	6 00	180 21.2	N 6 38.1	160 20.4	13.6	S 0 16.0	16.1	57.8
	01	195 21.4	37.2	174 53.0	13.5	0 32.1	16.1	57.9
	02	210 21.6	36.3	189 25.5	13.5	0 48.2	16.2	57.9
	03	225 21.8	.. 35.3	203 58.0	13.5	1 04.4	16.1	57.9
	04	240 22.0	34.4	218 30.5	13.5	1 20.5	16.1	57.9
	05	255 22.3	33.5	233 03.0	13.4	1 36.6	16.2	57.9
	06	270 22.5	N 6 32.5	247 35.4	13.4	S 1 52.8	16.1	58.0
	07	285 22.7	31.6	262 07.8	13.4	2 08.9	16.2	58.0
S	08	300 22.9	30.7	276 40.2	13.4	2 25.1	16.1	58.0
A	09	315 23.1	.. 29.8	291 12.6	13.3	2 41.2	16.2	58.0
T	10	330 23.3	28.8	305 44.9	13.3	2 57.4	16.1	58.0
U	11	345 23.5	27.9	320 17.2	13.2	3 13.5	16.2	58.1
R	12	0 23.7	N 6 27.0	334 49.4	13.2	S 3 29.7	16.1	58.1
D	13	15 23.9	26.0	349 21.6	13.2	3 45.8	16.1	58.1
A	14	30 24.1	25.1	3 53.8	13.2	4 01.9	16.1	58.1
Y	15	45 24.4	.. 24.2	18 26.0	13.1	4 18.0	16.1	58.1
	16	60 24.6	23.2	32 58.1	13.0	4 34.1	16.1	58.2
	17	75 24.8	22.3	47 30.1	13.1	4 50.2	16.1	58.2
	18	90 25.0	N 6 21.4	62 02.2	13.0	S 5 06.3	16.0	58.2
	19	105 25.2	20.4	76 34.2	12.9	5 22.3	16.1	58.2
	20	120 25.4	19.5	91 06.1	12.9	5 38.4	16.0	58.2
	21	135 25.6	.. 18.6	105 38.0	12.8	5 54.4	16.0	58.2
	22	150 25.8	17.6	120 09.8	12.9	6 10.4	16.0	58.3
	23	165 26.0	16.7	134 41.7	12.7	6 26.4	16.0	58.3
		S.D. 15.9	d 0.9	S.D. 15.5		15.7		15.8

Lat.	Twilight Naut.	Twilight Civil	Sunrise	Moonrise 4	Moonrise 5	Moonrise 6	Moonrise 7
°	h m	h m	h m	h m	h m	h m	h m
N 72	////	02 56	04 19	03 29	05 50	08 05	10 29
N 70	01 12	03 19	04 30	03 49	05 57	08 02	10 13
68	02 01	03 36	04 39	04 04	06 02	07 59	10 00
66	02 30	03 50	04 47	04 16	06 07	07 56	09 50
64	02 52	04 02	04 53	04 27	06 10	07 54	09 41
62	03 09	04 11	04 59	04 35	06 14	07 52	09 34
60	03 23	04 19	05 03	04 43	06 16	07 51	09 27
N 58	03 34	04 26	05 08	04 49	06 19	07 49	09 22
56	03 44	04 32	05 11	04 55	06 21	07 48	09 17
54	03 52	04 38	05 15	05 00	06 23	07 47	09 13
52	04 00	04 43	05 18	05 05	06 25	07 46	09 09
50	04 06	04 47	05 20	05 09	06 26	07 45	09 05
45	04 20	04 56	05 26	05 18	06 30	07 43	08 58
N 40	04 31	05 03	05 31	05 25	06 33	07 41	08 51
35	04 39	05 10	05 35	05 31	06 35	07 40	08 46
30	04 46	05 15	05 39	05 37	06 38	07 39	08 41
20	04 57	05 23	05 45	05 47	06 41	07 37	08 33
N 10	05 05	05 29	05 51	05 55	06 45	07 35	08 26
0	05 10	05 35	05 55	06 03	06 48	07 33	08 19
S 10	05 15	05 39	06 00	06 11	06 51	07 31	08 13
20	05 18	05 43	06 05	06 19	06 54	07 30	08 06
30	05 19	05 47	06 11	06 28	06 58	07 28	07 58
35	05 20	05 49	06 14	06 34	07 00	07 27	07 54
40	05 19	05 51	06 18	06 40	07 03	07 25	07 49
45	05 19	05 53	06 22	06 47	07 06	07 24	07 43
S 50	05 17	05 54	06 27	06 55	07 09	07 22	07 36
52	05 16	05 55	06 29	06 59	07 10	07 22	07 33
54	05 15	05 56	06 31	07 03	07 12	07 21	07 30
56	05 14	05 57	06 34	07 08	07 14	07 20	07 26
58	05 13	05 58	06 37	07 13	07 16	07 19	07 22
S 60	05 11	05 59	06 40	07 19	07 18	07 18	07 17

Lat.	Sunset	Twilight Civil	Twilight Naut.	Moonset 4	Moonset 5	Moonset 6	Moonset 7
°	h m	h m	h m	h m	h m	h m	h m
N 72	19 35	20 57	////	19 56	19 21	18 45	18 02
N 70	19 25	20 35	22 34	19 47	19 19	18 52	18 20
68	19 16	20 18	21 51	19 39	19 18	18 58	18 35
66	19 09	20 05	21 23	19 32	19 18	19 04	18 48
64	19 02	19 54	21 02	19 26	19 17	19 08	18 58
62	18 57	19 44	20 46	19 21	19 16	19 12	19 07
60	18 53	19 36	20 32	19 17	19 16	19 15	19 15
N 58	18 49	19 30	20 21	19 13	19 15	19 18	19 21
56	18 45	19 24	20 12	19 09	19 15	19 21	19 27
54	18 42	19 18	20 03	19 06	19 15	19 23	19 33
52	18 39	19 14	19 56	19 03	19 14	19 25	19 38
50	18 36	19 09	19 50	19 01	19 14	19 27	19 42
45	18 31	19 00	19 36	18 55	19 13	19 32	19 52
N 40	18 26	18 53	19 26	18 50	19 13	19 35	20 00
35	18 22	18 47	19 18	18 46	19 12	19 39	20 07
30	18 18	18 42	19 11	18 42	19 12	19 41	20 13
20	18 12	18 34	19 00	18 36	19 11	19 46	20 24
N 10	18 07	18 28	18 53	18 30	19 10	19 51	20 33
0	18 02	18 23	18 47	18 24	19 09	19 55	20 42
S 10	17 57	18 18	18 43	18 19	19 08	19 59	20 51
20	17 52	18 15	18 40	18 13	19 08	20 03	21 01
30	17 47	18 11	18 39	18 06	19 07	20 08	21 12
35	17 44	18 09	18 39	18 02	19 06	20 11	21 18
40	17 40	18 07	18 39	17 57	19 05	20 15	21 26
45	17 36	18 06	18 40	17 52	19 05	20 18	21 34
S 50	17 32	18 04	18 41	17 46	19 04	20 23	21 45
52	17 29	18 03	18 42	17 43	19 03	20 25	21 49
54	17 27	18 03	18 43	17 40	19 03	20 27	21 55
56	17 24	18 02	18 45	17 36	19 02	20 30	22 01
58	17 22	18 01	18 46	17 32	19 02	20 33	22 07
S 60	17 18	18 00	18 48	17 27	19 01	20 36	22 15

Day	SUN Eqn. of Time 00ʰ	SUN Eqn. of Time 12ʰ	SUN Mer. Pass.	MOON Mer. Pass. Upper	MOON Mer. Pass. Lower	MOON Age	MOON Phase
	m s	m s	h m	h m	h m	d	
4	00 45	00 55	11 59	12 14	24 36	00	●
5	01 05	01 14	11 59	12 59	00 36	01	
6	01 24	01 34	11 58	13 44	01 21	02	

FIGURE 8.2 ☆ *NAUTICAL ALMANAC*, RIGHT HAND DAILY PAGE (TYPICAL)

Go down the left hand column headed: Lat. to 35° N and 30° N. 33° N is close enough to our D. R. Lat. for interpolating purposes. An intermediate line of information for 33° N would be 3/5 of the difference of the times given for 35° N and 30° N.

Go across to the column headed: Twilight Nautical and obtain 1918 and 1911 respectively. Subtract 1911 from 1918, the difference is 7 minutes. Multiply 3/5 × 7 = 4.2 minutes or 4 minutes rounded. Add 4 minutes to 1911 obtaining and recording: 1915 TW NAUT as shown.

Go back one column to the column headed: Twilight Civil and obtain 1847 and 1842 respectively. Subtract 1842 from 1847, the difference is 5 minutes. Multiply 3/5 × 5 = 3 minutes. Add 3 minutes to 1842 and record: 1845 TW CIVIL, as shown.

Subtract 1845 from 1915 obtaining a working period of 30 minutes. Divide the working period by two obtaining 15 minutes, the middle of the working period. Add the 15 minutes to 1845 TW CIVIL and record: 1900 LOCAL TIME @ 135°.

By inspecting the log I estimated that we would be approaching the 129° meridian at twilight.

Go to the table at the beginning of the yellow pages headed: CONVERSION OF ARC TO TIME.

Go down the first column to 6°. The equivalent time for 6° is 24 minutes. Record 24 MINUS 6°. Subtract the 24 minutes from 1900 LOCAL TIME @ 135°, obtaining and recording: 1836 LOCAL TIME @ 129°.

Go to the third column in the CONVERSION OF ARC TO TIME table, and at the 135° full-hour meridian the conversion corresponds to 9 hours. Add 0900 to 1836; the result is 2736 SAME DAY GREEN. Subtract 2400 hours, obtaining and recording: 0336 GMT, 9/5/86. This would be the next day at Greenwich.

Advance the DR course line 98 nautical miles and establish the DR position at 1836 hours. Determine and record: D. R. Lat.: 32°49′ N; D. R. Long.: 128°56′ W; and Log: 425.

Use a plain sheet of paper for the next portion of the problem. As in Figure 8.3, label the paper as follows: STAR FINDER IDENTIFICATION FOR LOP #8, 9, & 10.

Record: 0336 GMT 9/5/86 USE TEMPLATE FOR 35° N.

Go to the left hand white pages in the almanac headed: 1986 SEPTEMBER 4, 5, 6 (THURS., FRI., SAT.) or Figure 7.2.

On the sheet of paper, calculate the LHA of Aries in the same manner as Polaris and record: LHA of Aries: 269°.

STAR FINDER IDENTIFICATION
FOR LOP #8, 9, & 10

0336 GMT 9-5-86 USE TEMPLATE FOR 35° N

GHA ♈ 3 hr	28° 55.7′
36 m	9° 01.5′
GHA ♈ 3hr 36m	37° 57′
	+360°
	397° 57′
LONG.	−128° 57′
LHA ♈	269°

GMT	LHA ♈	STAR	Hc	Zn
0332	268°	ALTAIR	53°	128°
0336	269°	ANTARES	26°	202°
0340	270°	ARCTURAS	38°	269°

FIGURE 8.3 ☆

I decided to sight three stars and assigned headings for GMT, LHA of Aries, Star, Hc, and Zn.

Go down the GMT column and record: 0332, 0336, and 0340. The times of these sights were selected as close as possible to the middle of the working period. Recall that 0336 GMT represents the middle of the working period and that the LHA of Aries corresponding to this time is 269°.

The earth rotates 1° in every 4 minutes. If you forget that the earth rotates 1° in every 4′, consult the table: CONVERSION OF ARC TO TIME.

Go down the LHA of Aries column and record: 268°, 269°, and 270° corresponding to the convenient 4-minute intervals.

Take out the star finder. One of the uses of the 2102-D Star

Finder and Identifier is to determine the approximate altitude and azimuth of 57 stars especially selected for navigational purposes. These are the same stars that are found in the INDEX TO SELECTED STARS printed on one side of the white card insert and toward the end of the yellow pages in the almanac. The names of these stars are printed on both sides of the circular base of the star finder.

One side of the base is labelled N and the other S. If sailing in North Latitude use the side labelled N, and if sailing in South Latitude use the side labelled S.

The positions of the stars are all identified by a dot, a circle around the dot, and followed by the name of the star. The circles represent the magnitude of the stars. The largest heavy ring represents first magnitude, the next smaller ring represents second magnitude, and a yet smaller and thinner ring represents third magnitude. The brightest stars are of the first magnitude.

The accuracy of the device is approximately 5° for both altitude and azimuth.

Ten templates are provided. Nine of the templates are to be used for every 10° of latitude, north or south.

The template closest to our DR is labelled 35° N. The template is attached to the N side of the base by a small pin in the center of the base.

The LHA is printed on the base every 5° in ½° gradations. Rotate the template until the arrow on the template corresponds to 268° on the base. 268° is the LHA of Aries of the first sight.

Only a percentage of the stars are visible within the blue grid. The outer portion of the grid represents the observer's horizon, and the center the zenith, that is, directly overhead. The blue lines extending radially from the center outward are for determining azimuth and the blue continuous lines around the center are for determining altitude. The blue lines are spaced at 5° intervals.

Your task is to select three stars of sufficient magnitude and with good cuts or angles between their azimuths. Because I had initially intended to use sight reduction tables limited to declinations between 0° and 29° for the course, I selected stars within these limits.

Open the almanac to the white left hand page dated 9/4/86 or go to Figure 7.2.

Go to the last column on the right side of the page headed: STARS and use the declination information in conjunction with the star finder selections.

Note that there are six first magnitude stars visible at 35° N and LHA of Aries-268°. Deneb and Vega were rejected because their declinations were greater than 29° (not necessary to reject with the sight reduction tables in the almanac). Spica's altitude is only 10°, and its azimuth is within 20° of Arcturus. Arcturus has a higher altitude and declination less than 29°. Altair, Antares, and Arcturus were selected. Their azimuths provided reasonable angular separation for plotting purposes.

Go down the column you headed: STARS and record: ALTAIR, ANTARES, and ARCTURUS. These stars were used for observations in order of increasing azimuth because the horizon would disappear first in the East.

It is important to realize that it isn't always possible to obtain three first magnitude stars that will yield good angular separation.

As evening twilight deepens, the brightest stars are seen first, so in the evening, observations are made on stars of the greatest magnitude first. At morning twilight, altitude measurements are taken first on stars of lesser magnitute, as they fade from sight in the growing light while brighter stars can still be seen.

Another consideration in selecting stars is the path of the moon. I once made the error of selecting a star that was close to the moon and could not see the star at the appropriate time.

The altitude of Altair appears to be between 50° and 55°. I estimated and recorded: Hc: 53°.

The dot appears to be between azimuths 125° and 130° and is recorded: Zn: 128°

Move the arrow to the 269° graduation on the base. The next star going from East to West would be Antares, and finally, at LHA Aries of 270°, Arcturus, almost due West.

Find the altitudes and azimuths of the last two stars and check your results against the list.

You have all the information required to sight these stars. You have the time of each sight, the altitude to preset the sextant, and the azimuth (direction toward the star).

It is important to know that the star finder is not a direct chart of stars in the sky, and it is not possible to use it for viewing their relative positions in the sky.

Continue to complete the worksheet for LOP #8 and record: Body: ALTAIR and Local Time: 1832. Begin similar worksheets recording: No.: 9; Area: NORTH PACIFIC; Body: ANTARES; Local Time: 1836 and No.: 10; Area: NORTH PACIFIC; Body: ARCTURUS; Local Time: 1840.

The DR information and log readings are all the same because the sights were taken within minutes of each other. . . . See Figures 8.4 and 8.5.

I wasn't paying attention to the time, so it wasn't until 1831 local time, or 0331 GMT, that the receiver was tuned to stations WWV and WWVH.

At 3h 32m a stopwatch was started, the sextant was preset at 53°, and at an azimuth of approximately 128°, determined by sighting over the binnacle and visually correcting for magnetic variation, the star Altair was seen in the sextant horizon glass lying slightly above the horizon.

The mate let me know that the time for a two-minute-late sight was approaching, and I yelled "mark" about six seconds later than the moment when Altair touched the horizon. My excuse, if I need one, is that I got a late start.

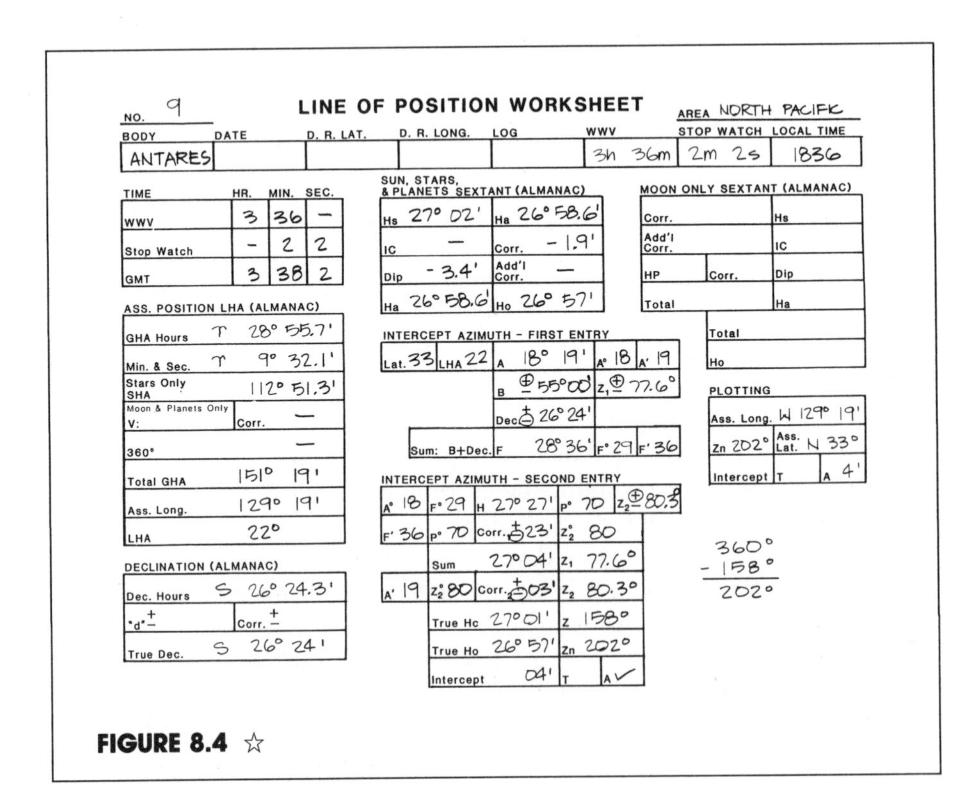

LINE OF POSITION WORKSHEET

NO. 9 — AREA NORTH PACIFIC

BODY	DATE	D. R. LAT.	D. R. LONG.	LOG	WWV	STOP WATCH	LOCAL TIME
ANTARES					3h 36m	2m 2s	1836

TIME	HR.	MIN.	SEC.
WWV	3	36	—
Stop Watch	-	2	2
GMT	3	38	2

SUN, STARS, & PLANETS SEXTANT (ALMANAC)

Hs	27° 02'	Ha	26° 58.6'
IC	—	Corr.	- 1.9'
Dip	- 3.4'	Add'l Corr.	—
Ha	26° 58.6'	Ho	26° 57'

MOON ONLY SEXTANT (ALMANAC)

Corr.		Hs
Add'l Corr.		IC
HP	Corr.	Dip
Total		Ha
	Total	
	Ho	

ASS. POSITION LHA (ALMANAC)

GHA Hours	♈	28° 55.7'
Min. & Sec.	♈	9° 32.1'
Stars Only SHA		112° 51.3'
Moon & Planets Only V:	Corr.	—
360°		—
Total GHA		151° 19'
Ass. Long.		129° 19'
LHA		22°

INTERCEPT AZIMUTH – FIRST ENTRY

Lat. 33	LHA 22	A 18° 19'	A° 18	A' 19
		B ⊕55°00'	Z₁ ⊕ 77.6°	
		Dec ± 26° 24'		
	Sum: B+Dec.	F 28° 36'	F° 29	F' 36

PLOTTING

Ass. Long.	W 129° 19'
Zn 202°	Ass. Lat. N 33°
Intercept T	A 4'

INTERCEPT AZIMUTH – SECOND ENTRY

A° 18	F° 29	H 27° 27'	P° 70	Z₂ ⊕ 80.3
F' 36	P° 70	Corr. ± 23'	Z₂° 80	
	Sum	27° 04'	Z₁ 77.6°	
A' 19	Z₂° 80	Corr. ± 03'	Z₂ 80.3°	
	True Hc	27° 01'	Z 158°	
	True Ho	26° 57'	Zn 202°	
	Intercept	04'	T	A ✓

DECLINATION (ALMANAC)

Dec. Hours		S 26° 24.3'
"d" ±	Corr. ±	
True Dec.		S 26° 24'

360°
− 158°
202°

FIGURE 8.4 ☆

Anyway, the time and sextant readings were recorded: Stop Watch: 2m 6s and Hs: 54°06′.

With not much time to spare, a stopwatch was started at 3h 36m. The sextant was preset at 26°. I looked in the direction of 202° and took the Antares sight. The stopwatch was stopped and the readings recorded: Stop Watch: 2m 2s and Hs: 27°02′.

The final sight, that of Arcturus, was accomplished in a similar manner and recorded: WWV: 3h 40m; Stop Watch: 2m 0s; and Hs: 37°55′.

Go to the almanac left hand white page for 9/5/86 or Figure 7.2.

Thus far we have been determining the LHA of Aries. For these sights we will be determining the LHA of the individual stars. The big difference is that we will be applying the *sidereal hour angle* (SHA) for each star. The SHA of each star is nothing more than

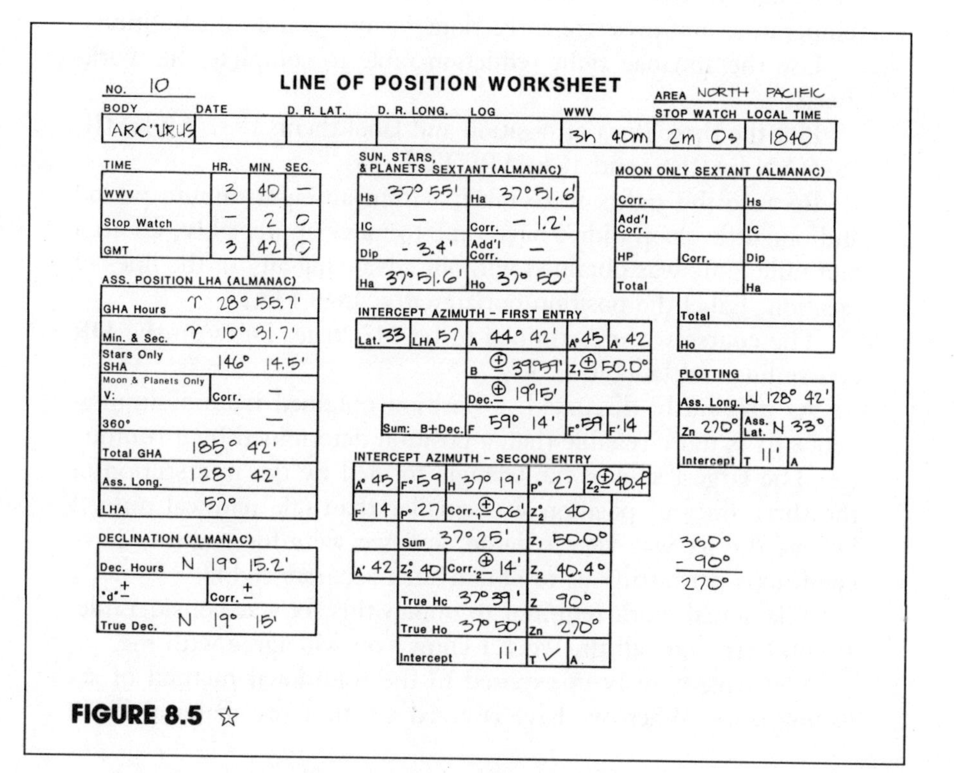

NO. 10

LINE OF POSITION WORKSHEET

AREA NORTH PACIFIC

BODY	DATE	D. R. LAT.	D. R. LONG.	LOG	WWV	STOP WATCH	LOCAL TIME
ARC'TURUS					3h 40m	2m 0s	1840

TIME	HR.	MIN.	SEC.
WWV	3	40	—
Stop Watch	—	2	0
GMT	3	42	0

ASS. POSITION LHA (ALMANAC)

GHA Hours	♈ 28° 55.7′
Min. & Sec.	♈ 10° 31.7′
Stars Only SHA	146° 14.5′
Moon & Planets Only V:	Corr. —
360°	—
Total GHA	185° 42′
Ass. Long.	128° 42′
LHA	57°

DECLINATION (ALMANAC)

Dec. Hours	N 19° 15.2′
"d" ±	Corr. ±
True Dec.	N 19° 15′

SUN, STARS, & PLANETS SEXTANT (ALMANAC)

Hs 37° 55′	Ha 37° 51.6′
IC —	Corr. − 1.2′
Dip − 3.4′	Add'l Corr. —
Ha 37° 51.6′	Ho 37° 50′

INTERCEPT AZIMUTH – FIRST ENTRY

Lat. 33	LHA 57	A 44° 42′	A° 45	A′ 42
		B ⊕ 39° 59′	Z_1 ⊕ 50.0°	
		Dec. ⊕ 19° 15′		
	Sum: B+Dec.	F 59° 14′	F° 59	F′ 14

INTERCEPT AZIMUTH – SECOND ENTRY

A° 45	F° 59	H 37° 19′	P° 27	Z_2 ⊕ 40.4°
F′ 14	P° 27	$Corr._1$ ⊕ 06′	$Z_2°$ 40	
	Sum 37° 25′		Z_1 50.0°	
A′ 42	$Z_2°$ 40	$Corr._2$ ⊕ 14′	Z_2 40.4°	
	True Hc 37° 39′		Z 90°	
	True Ho 37° 50′		Zn 270°	
	Intercept 11′		T ✓	A

MOON ONLY SEXTANT (ALMANAC)

Corr.		Hs
Add'l Corr.		IC
HP	Corr.	Dip
Total		Ha
Total		
Ho		

PLOTTING

Ass. Long. W 128° 42′	
Zn 270°	Ass. Lat. N 33°
Intercept T 11′	A

360°
− 90°
270°

FIGURE 8.5 ☆

the angle that the star makes with the GHA of Aries and is always positive and added.

The tables for all of the 57 selected navigational stars are reduced to a tidy little column of information located on the right hand side of the left hand white pages headed: STARS.

Beginning with Altair, record: SHA: 62°27.9′ in the space provided on the worksheet. Record the declination in the space: Dec. Hours: N 8°50.0′ and round off and record: True Dec.: N 8°50′.

Unlike the sun there are no little d corrections, because the stars do not change declination appreciably over a short period of time as do the sun, moon, and planets.

In the same manner, record the sidereal hour angle angles and declinations of Antares and Arcturus.

Continuing with the almanac, record the GHA Hours of Aries, the Min. & Sec. of Aries, and apply the Ass. Long. to the Total GHA to obtain the LHA of each sight.

Make corrections to the sextant altitudes. There wasn't any index error, the height of my eye above the water was 11 feet for Altair and 12 feet for Antares and Arcturus, and there were no temperature and pressure corrections for non-standard conditions.

Use the almanac sight reduction table to complete the worksheets.

Plot the three lines of position and label them: 1832/ALTAIR, 1836/ANTARES, and 1840/ARCTURUS.

Because the sights were taken within an eight-minute period and our little sloop didn't have time to move appreciably, about a half mile, a fix was obtained without advancing any of the lines of position. Label the position horizontally: 1840 FIX.

The course was determined to be 247° true. Draw in the DR course line and label it: C247.

As previously discussed, a position obtained from a simultaneous fix is more reliable than a position determined by a running fix. The largest side of the triangle formed by the intersection of the three lines of position was smaller than one nautical mile. I believe the fix was very reliable, and we were fortunate to have conditions permitting us to obtain such accurate sights.

The actual work required to obtain this fix was considerable. If you have done all the work I know you will agree with me.

You have now been exposed to the traditional method of selecting stars. When we have covered a little more material I will show you a faster way to select stars, reduce sights, and obtain a fix. This will be accomplished by using *Vol. I, Pub. No. 249 Sight Reduction Tables for Air Navigation*. With Vol. I you will be able to use almost all of the 57 selected stars (but not all at any one place on any given day).

DOLPHINS, NORTH PACIFIC

NINE

MOON AND PLANET SIGHTS

Seventeen days have passed and it is Sunday, 9/21/86. We have been at sea for twenty-one days. We have been sailing in the trade winds for a few days now and although the trades have been reasonably light, we've been averaging better than 100 nautical miles per day.

We probably would have completed our voyage by now if we had left in the spring or summer months when winds are usually stronger.

During the last three weeks we have caught many beautiful

fish—recently Mahi Mahi and earlier, in colder waters, Albacore tuna. We have been entertained royally by large schools of dolphins. They swim alongside our yacht, jumping and occasionally spinning for us. Dolphins like to swim under the bow and we have observed them for hours from the foredeck. After awhile they would depart, swimming at high speed and jumping over large distances. My guess is that the pods were responding to sightings of fish schools.

We didn't navigate during the daylight hours on 9/21/86 and decided to take observations on the end of the full moon and Jupiter.

One advantage to using the moon is that the moon is often available for an additional sight at unusual hours.

Begin a worksheet, as shown in Figure 9.1, recording as follows: No.: 44; Area: NORTH PACIFIC; Body: MOON; and Date: 9/21/86.

Calculating sunset twilight for this observation will not be necessary because the moonlit horizon will be visible most of the night. However, it is advantageous to know when the moon will become visible and at sufficient altitude for obtaining a sight.

Open the almanac to the right hand white page headed: 1986 SEPTEMBER 19, 20, 21 (FRI., SAT., SUN.) or go to Figure 9.2.

If the DR course line was extended to 2200, we would be sailing a little below 24° N and a little west of 151° W as shown in Figure 9.3.

Go to the upper right hand portion of the page to the column headed: Lat.

Go down the Lat. column to 30° N and 20° N.

Go across to the column headed: Moonrise, and stop at the column headed: 21. The number 21 represents the date. Obtain 2006 and 1951 hours.

In the space for calculations on the worksheet record: 2006 and 1951 hours as shown. Subtract 1951 from 2006 hours and obtain a difference of 15 minutes. N 24° is ⅖ of the difference in latitude. Multiply ⅖ × 15 = 6 minutes less than 2006 hours or 2000 hours at 150°, the full-hour meridian. Record: 2000 @ 150°.

We will be sailing at approximately 151° W or 1° = 4 minutes. Add 4 minutes to 2000 hours, obtaining and recording: 2004 MOONRISE @ 151°. To allow the moon some time to rise well into the sky, advance the course line to 2200 as shown in Figure 9.3.

The log ceased operating about two weeks ago and we have been gaining expertise in estimating our DR position. I obtained and recorded: D. R. Long.: 151°04 W and D.R. Lat.: 23°42′ N.

NO. 44 **LINE OF POSITION WORKSHEET** AREA NORTH PACIFIC

BODY	DATE	D. R. LAT.	D. R. LONG.	LOG	WWV	STOP WATCH	LOCAL TIME
MOON	9-21-86	N 23° 42'	W 151° 04'	—	7h 56m	2m 38s	2200

TIME	HR.	MIN.	SEC.
WWV	7	56	—
Stop Watch	—	2	38
GMT	7	58	38

ASS. POSITION LHA (ALMANAC)

GHA Hours	62° 13.4'	
Min. & Sec.	13° 59.4'	
Stars Only SHA	—	
Moon & Planets Only v: 13.3	Corr. 13.0'	
360°	360°	
Total GHA	436° 26'	
Ass. Long.	151° 26'	
LHA	285°	

DECLINATION (ALMANAC)

Dec. Hours	N 19° 03.1'
"d" ⊕ 11.1	Corr. ⊕ 10.8'
True Dec.	N 19° 14'

SUN, STARS, & PLANETS SEXTANT (ALMANAC)

Hs		Ha	
IC		Corr.	
Dip		Add'l Corr.	
Ha		Ho	

MOON ONLY SEXTANT (ALMANAC)

Corr.	62.1'	Hs	20° 07'
Add'l Corr.	0.1'	IC	2'
HP 55.0	Corr. 2.0'	Dip	−3.1'
Total	64.2'	Ha	20° 05.9'
		Total	1° 04.2'
		Ho	21° 10'

INTERCEPT AZIMUTH – FIRST ENTRY

Lat. 24	LHA 285	A 61° 56'	A° 62	A' 56
		B ⊕ 30° 10'	Z_1 ⊕ 33.4°	
		Dec. ⊕ 19° 14'		
Sum: B+Dec.		F 49° 24'	F° 49	F' 24

PLOTTING

Ass. Long.	W 151° 26'
Zn 78°	Ass. Lat. N 24°
Intercept	T 14' A

INTERCEPT AZIMUTH – SECOND ENTRY

A° 62	F° 49	H 20° 45'	P° 19
F' 24	P° 19	$Corr._1$ ⊕ 08'	$Z_2°$ 45
		Sum 20° 53'	Z_1 33.4°
A' 56	$Z_2°$ 45	$Corr._2$ ⊕ 03'	Z_2 44.6°
		True Hc 20° 56'	Z 78°
		True Ho 21° 10'	Zn 78°
		Intercept 14'	T ✓ A

Z_2 ⊕ 44.6

2200
1000 GMT @ 151°
3200 SAME DAY
(2400) GREENWICH
0800 GMT 9/22/86

2006
(1951)
15

2006
(6)
2000 @ 150°
4
2004 MOONRISE @ 151°

FIGURE 9.1 ☆

There is some controversy in taking sights of the moon and other bodies at night. Moonglow can affect the horizon, causing it to appear to be below the true horizon and increasing the sextant altitude. The same effect is experienced when sighting a star or planet too close in azimuth to the moonglow on the horizon. For our level of accuracy this error will be neglected.

At 150° the difference between the local time and GMT is 10 hours. Add 1000 GMT © 150° to 2200, obtain and record: 3200 SAME DAY @ GREENWICH.

Subtract 2400 hours and record: 0800 GMT, 9/22/86. This is the next day at Greenwich.

A stopwatch was started at: WWV: 7h 56m.

Firmly attached to the port side stern rail via my safety harness, I proceeded to measure the altitude of the lower limb of the moon.

For purposes of this example, we will say that the sextant was checked for index error by setting it on 0°00′ and viewing the

1986 SEPTEMBER 19, 20, 21 (FRI., SAT., SUN.)

G.M.T. (UT)		SUN G.H.A.	SUN Dec.	MOON G.H.A.	MOON *v*	MOON Dec.	MOON *d*	MOON H.P.
d h		° ′	° ′	° ′	′	° ′	′	′
19 00		181 29.6	N 1 41.0	352 16.7	14.5	N 0 45.3	15.6	56.9
01		196 29.8	40.0	6 50.2	14.5	1 00.9	15.6	56.9
02		211 30.0	39.0	21 23.7	14.6	1 16.5	15.6	56.9
03		226 30.2	.. 38.1	35 57.3	14.6	1 32.1	15.6	56.9
04		241 30.4	37.1	50 30.9	14.5	1 47.7	15.6	56.8
05		256 30.7	36.1	65 04.4	14.7	2 03.3	15.5	56.8
06		271 30.9	N 1 35.2	79 38.1	14.6	N 2 18.8	15.5	56.8
07		286 31.1	34.2	94 11.7	14.6	2 34.3	15.5	56.8
08		301 31.3	33.2	108 45.3	14.7	2 49.8	15.4	56.7
09	FRIDAY	316 31.6	.. 32.3	123 19.0	14.6	3 05.2	15.5	56.7
10		331 31.8	31.3	137 52.6	14.7	3 20.7	15.4	56.7
11		346 32.0	30.3	152 26.3	14.7	3 36.1	15.3	56.7
12		1 32.2	N 1 29.4	167 00.0	14.7	N 3 51.4	15.4	56.6
13		16 32.5	28.4	181 33.7	14.7	4 06.8	15.3	56.6
14		31 32.7	27.4	196 07.4	14.7	4 22.1	15.3	56.6
15		46 32.9	.. 26.4	210 41.1	14.8	4 37.4	15.2	56.5
16		61 33.1	25.5	225 14.9	14.7	4 52.6	15.2	56.5
17		76 33.3	24.5	239 48.6	14.8	5 07.8	15.2	56.5
18		91 33.6	N 1 23.5	254 22.4	14.7	N 5 23.0	15.1	56.5
19		106 33.8	22.6	268 56.1	14.7	5 38.1	15.1	56.4
20		121 34.0	21.6	283 29.8	14.8	5 53.2	15.1	56.4
21		136 34.2	.. 20.6	298 03.6	14.7	6 08.3	15.0	56.4
22		151 34.5	19.7	312 37.3	14.8	6 23.3	15.0	56.4
23		166 34.7	18.7	327 11.1	14.7	6 38.3	14.9	56.3
20 00		181 34.9	N 1 17.7	341 44.8	14.8	N 6 53.2	14.9	56.3
01		196 35.1	16.8	356 18.6	14.7	7 08.1	14.8	56.3
02		211 35.4	15.8	10 52.3	14.8	7 22.9	14.8	56.3
03		226 35.6	.. 14.8	25 26.1	14.7	7 37.7	14.8	56.2
04		241 35.8	13.8	39 59.8	14.7	7 52.5	14.7	56.2
05		256 36.0	12.9	54 33.5	14.8	8 07.2	14.7	56.2
06		271 36.2	N 1 11.9	69 07.3	14.7	N 8 21.9	14.6	56.2
07		286 36.5	10.9	83 41.0	14.7	8 36.5	14.5	56.1
08	SATURDAY	301 36.7	10.0	98 14.7	14.7	8 51.0	14.5	56.1
09		316 36.9	.. 09.0	112 48.4	14.7	9 05.5	14.5	56.1
10		331 37.1	08.0	127 22.1	14.7	9 20.0	14.4	56.1
11		346 37.4	07.0	141 55.8	14.6	9 34.4	14.4	56.0
12		1 37.6	N 1 06.1	156 29.4	14.7	N 9 48.8	14.3	56.0
13		16 37.8	05.1	171 03.1	14.6	10 03.1	14.2	56.0
14		31 38.0	04.1	185 36.7	14.6	10 17.3	14.2	56.0
15		46 38.2	.. 03.2	200 10.3	14.6	10 31.5	14.1	55.9
16		61 38.5	02.2	214 43.9	14.6	10 45.6	14.1	55.9
17		76 38.7	01.2	229 17.5	14.6	10 59.7	14.0	55.9
18		91 38.9	N 1 00.3	243 51.1	14.5	N11 13.7	14.0	55.9
19		106 39.1	0 59.3	258 24.6	14.6	11 27.7	13.9	55.8
20		121 39.4	58.3	272 58.2	14.5	11 41.6	13.8	55.8
21		136 39.6	.. 57.3	287 31.7	14.5	11 55.4	13.8	55.8
22		151 39.8	56.4	302 05.2	14.4	12 09.2	13.7	55.8
23		166 40.0	55.4	316 38.6	14.5	12 22.9	13.6	55.7
21 00		181 40.3	N 0 54.4	331 12.1	14.4	N12 36.5	13.6	55.7
01		196 40.5	53.5	345 45.5	14.4	12 50.1	13.5	55.7
02		211 40.7	52.5	0 18.9	14.4	13 03.6	13.5	55.7
03		226 40.9	.. 51.5	14 52.3	14.3	13 17.1	13.3	55.6
04		241 41.1	50.5	29 25.6	14.3	13 30.4	13.3	55.6
05		256 41.4	49.6	43 58.9	14.3	13 43.7	13.3	55.6
06		271 41.6	N 0 48.6	58 32.2	14.3	N13 57.0	13.2	55.6
07		286 41.8	47.6	73 05.5	14.2	14 10.2	13.1	55.5
08		301 42.0	46.7	87 38.7	14.2	14 23.3	13.0	55.5
09	SUNDAY	316 42.2	.. 45.7	102 11.9	14.2	14 36.3	12.9	55.5
10		331 42.5	44.7	116 45.1	14.1	14 49.2	12.9	55.5
11		346 42.7	43.7	131 18.2	14.1	15 02.1	12.8	55.4
12		1 42.9	N 0 42.8	145 51.3	14.1	N15 14.9	12.8	55.4
13		16 43.1	41.8	160 24.4	14.0	15 27.7	12.6	55.4
14		31 43.4	40.8	174 57.4	14.0	15 40.3	12.6	55.4
15		46 43.6	.. 39.9	189 30.4	14.0	15 52.9	12.5	55.4
16		61 43.8	38.9	204 03.4	13.9	16 05.4	12.4	55.3
17		76 44.0	37.9	218 36.3	13.9	16 17.8	12.4	55.3
18		91 44.2	N 0 36.9	233 09.2	13.9	N16 30.2	12.2	55.3
19		106 44.5	36.0	247 42.1	13.8	16 42.4	12.2	55.3
20		121 44.7	35.0	262 14.9	13.8	16 54.6	12.1	55.2
21		136 44.9	.. 34.0	276 47.7	13.8	17 06.7	12.0	55.2
22		151 45.1	33.1	291 20.5	13.7	17 18.7	12.0	55.2
23		166 45.4	32.1	305 53.2	13.6	17 30.7	11.8	55.2
		S.D. 16.0	*d* 1.0	S.D.	15.4		15.3	15.1

Lat.	Twilight Naut.	Twilight Civil	Sunrise	Moonrise 19	Moonrise 20	Moonrise 21	Moonrise 22
°	h m	h m	h m	h m	h m	h m	h m
N 72	02 43	04 18	05 28	17 57	17 17	16 15	▭
N 70	03 06	04 28	05 31	18 04	17 35	16 54	▭
68	03 24	04 36	05 33	18 10	17 48	17 21	16 33
66	03 37	04 43	05 35	18 15	18 00	17 42	17 16
64	03 49	04 48	05 36	18 19	18 10	17 59	17 45
62	03 58	04 53	05 37	18 23	18 18	18 13	18 08
60	04 06	04 57	05 39	18 26	18 25	18 25	18 26
N 58	04 13	05 00	05 40	18 29	18 31	18 35	18 41
56	04 18	05 03	05 40	18 31	18 37	18 44	18 54
54	04 23	05 06	05 41	18 34	18 42	18 52	19 05
52	04 28	05 08	05 42	18 36	18 47	19 00	19 16
50	04 32	05 10	05 43	18 38	18 51	19 06	19 25
45	04 40	05 15	05 44	18 42	19 00	19 20	19 44
N 40	04 46	05 18	05 45	18 46	19 08	19 32	19 59
35	04 51	05 21	05 46	18 49	19 15	19 42	20 13
30	04 55	05 23	05 47	18 52	19 20	19 51	20 24
20	05 01	05 26	05 48	18 56	19 31	20 06	20 44
N 10	05 04	05 28	05 49	19 01	19 40	20 20	21 02
0	05 06	05 30	05 50	19 05	19 48	20 32	21 18
S 10	05 06	05 30	05 51	19 09	19 57	20 45	21 35
20	05 04	05 30	05 52	19 13	20 06	20 59	21 53
30	05 01	05 29	05 53	19 19	20 17	21 15	22 13
35	04 58	05 28	05 53	19 22	20 23	21 24	22 25
40	04 55	05 26	05 53	19 25	20 30	21 35	22 39
45	04 50	05 24	05 54	19 29	20 38	21 47	22 56
S 50	04 44	05 22	05 54	19 34	20 48	22 03	23 17
52	04 41	05 21	05 54	19 36	20 53	22 10	23 27
54	04 38	05 19	05 55	19 38	20 58	22 18	23 38
56	04 34	05 18	05 55	19 41	21 04	22 27	23 50
58	04 30	05 16	05 55	19 44	21 10	22 37	24 05
S 60	04 25	05 14	05 55	19 47	21 18	22 49	24 23

Lat.	Sunset	Twilight Civil	Twilight Naut.	Moonset 19	Moonset 20	Moonset 21	Moonset 22
°	h m	h m	h m	h m	h m	h m	h m
N 72	18 17	19 25	20 59	07 15	09 27	12 02	▭
N 70	18 14	19 16	20 37	07 11	09 13	11 25	▭
68	18 12	19 08	20 20	07 08	09 01	11 00	13 23
66	18 11	19 02	20 07	07 06	08 51	10 40	12 41
64	18 09	18 57	19 56	07 03	08 43	10 25	12 12
62	18 08	18 53	19 47	07 02	08 36	10 12	11 51
60	18 07	18 49	19 39	07 00	08 30	10 01	11 34
N 58	18 06	18 45	19 33	06 59	08 25	09 52	11 19
56	18 05	18 43	19 27	06 57	08 21	09 44	11 07
54	18 05	18 40	19 22	06 56	08 17	09 36	10 56
52	18 04	18 38	19 18	06 55	08 13	09 30	10 47
50	18 03	18 36	19 14	06 54	08 10	09 24	10 38
45	18 02	18 31	19 06	06 52	08 02	09 11	10 20
N 40	18 01	18 28	19 00	06 51	07 56	09 01	10 06
35	18 00	18 26	18 55	06 49	07 51	08 52	09 54
30	18 00	18 23	18 51	06 48	07 47	08 45	09 43
20	17 58	18 20	18 46	06 46	07 39	08 31	09 25
N 10	17 58	18 18	18 43	06 44	07 32	08 20	09 09
0	17 57	18 17	18 41	06 42	07 26	08 09	08 54
S 10	17 56	18 17	18 41	06 40	07 19	07 59	08 40
20	17 55	18 17	18 43	06 38	07 12	07 47	08 24
30	17 55	18 19	18 47	06 36	07 05	07 34	08 06
35	17 55	18 20	18 49	06 35	07 00	07 27	07 56
40	17 54	18 21	18 53	06 34	06 55	07 18	07 44
45	17 54	18 23	18 58	06 32	06 50	07 09	07 30
S 50	17 54	18 26	19 04	06 30	06 43	06 57	07 13
52	17 54	18 27	19 07	06 29	06 40	06 51	07 05
54	17 54	18 29	19 10	06 28	06 36	06 45	06 57
56	17 53	18 31	19 14	06 27	06 32	06 39	06 47
58	17 53	18 32	19 19	06 26	06 28	06 31	06 36
S 60	17 53	18 35	19 24	06 24	06 24	06 23	06 23

Day	SUN Eqn. of Time 00ʰ	SUN Eqn. of Time 12ʰ	SUN Mer. Pass.	MOON Mer. Pass. Upper	MOON Mer. Pass. Lower	MOON Age	MOON Phase
	m s	m s	h m	h m	h m	d	
19	05 58	06 08	11 54	00 32	12 54	15	○
20	06 19	06 30	11 54	01 15	13 37	16	
21	06 41	06 51	11 53	01 59	14 21	17	

FIGURE 9.2 ☆ *NAUTICAL ALMANAC*, RIGHT HAND DAILY PAGE (TYPICAL)

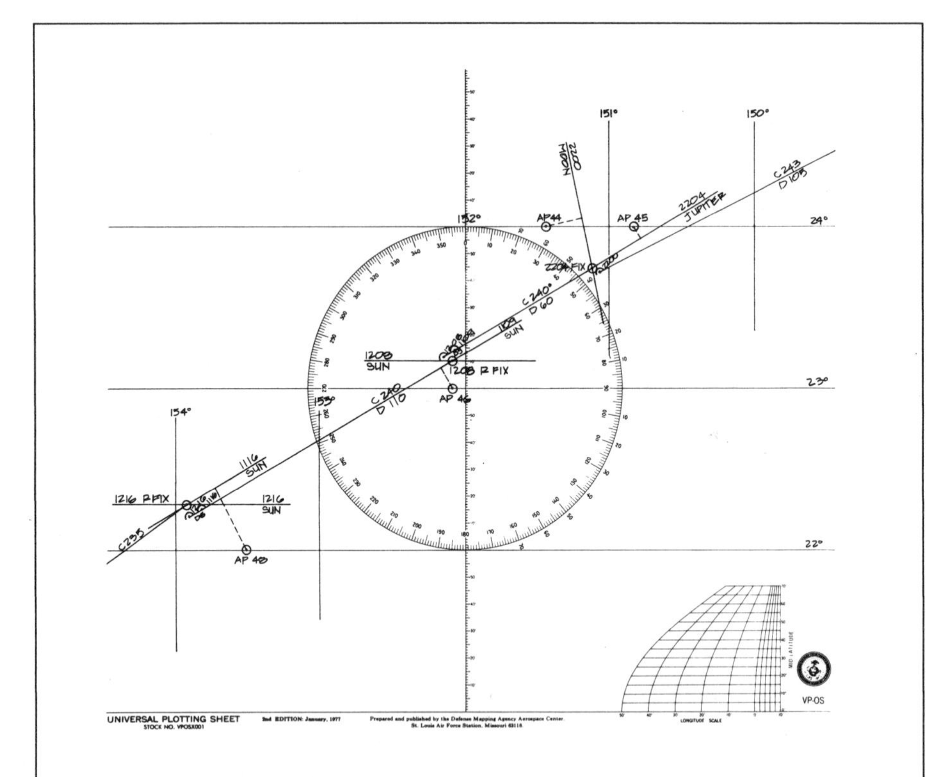

FIGURE 9.3 ☆ DEAD RECKONING AND TERRESTRIAL POSITIONS FROM 9/21/86 TO 9/23/86

horizon. The micrometer drum was rotated slightly to view a continuous horizon. The mark on the index arm corresponding to the micrometer dial read 58′. It is best to check for this error in daylight.

Without changing the settings on the sextant I continued with the sight. Except in the case where there is excessive glare on the horizon, it is usually not necessary to use shades. I have occasionally used a little horizon shading to eliminate excessive moonglow.

Another sighting technique is to view the moon directly through the telescope, depress the release lever, and follow the body to the horizon, while moving the index arm forward.

When the moon appears to be near the horizon, let go of the release lever, rotate the micrometer drum until the lower limb of the moon just touches the horizon, and time the sight.

Having taken the moon sight, the elapsed time on the stopwatch was recorded: Stop Watch: 2m 38s.

Record: Local Time: 2200.

In the space provided in the block: Moon Only Sextant (Almanac), record the sextant altitude: Hs: 20°07′.

Record the index error: IC: 2′.

Add the times to obtain: GMT: 7h 58m 38s.

Open the almanac to the right hand white page headed: 1986 SEPTEMBER 22, 23, 24(MON., TUES., WED.) or go to Figure 9.4 and go down the G.M.T. column to MONDAY the 22nd. to 07 h.

Go across to the column headed: MOON. In addition to G.H.A. and Dec., note that there are three additional corrections for the moon. Realize that the path of the moon around the earth is the most complex of any body you will be using. The nice thing about applying these simple corrections is that they adjust for the complex path of the moon.

Obtain and record as follows: GHA: 62°13.4′; v: 13.3 (in the Moon & Planets Only space); Dec. Hours: 19°03.1′ N; "d" ±: 11.1 circling the + sign because declination is increasing; and HP: 55.0 (in the space provided in the Moon Only Sextant block).

Go to the inside back cover of your current year almanac. The last page and the inside back cover contain information for correcting the altitude of moon sights.

Go to the last page headed: ALTITUDE CORRECTION TABLES 0°–35°—MOON.

Go to the block headed: DIP. My eye was about 10 feet above water. Obtain and record: Dip: − 3.1′. Add the dip and index error corrections algebraically obtaining and recording: Ha: 20°05.9′.

Go to the column headed: App. Alt. 20°–24°.

Go down the column to the nearest minute . . . 10′ is close enough. Obtain and record: Corr.: 62.1′.

Go down the same column to the columns headed: L and U. Select the column headed: L for lower limb . . . go down the L column and the H. P. column to 55.2. 55.2 is the closest value to 55.0, the HP from the almanac. The corresponding correction under L is obtained and recorded: Corr.: 2.0′.

If the upper limb of the moon had been sighted, 30′ would have been subtracted from the apparent altitude and the upper limb correction: U would have been added in the same manner as the lower limb corr.: L.

The temperature was 75°F and the pressure was 1040mb. Therefore, there were no corrections for non-standard conditions.

Add the corrections, obtaining and recording: Total: 64.2′. 64.2′

1986 SEPTEMBER 22, 23, 24 (MON., TUES., WED.) 187

G.M.T. (UT) d h	SUN G.H.A. ° ′	SUN Dec. ° ′	MOON G.H.A. ° ′	MOON v ′	MOON Dec. ° ′	MOON d ′	MOON H.P. ′
22 00 (MONDAY)	181 45.6	N 0 31.1	320 25.8	13.7	N17 42.5	11.8	55.2
01	196 45.8	30.1	334 58.5	13.5	17 54.3	11.7	55.1
02	211 46.0	29.2	349 31.0	13.6	18 06.0	11.6	55.1
03	226 46.2	· · 28.2	4 03.6	13.5	18 17.6	11.5	55.1
04	241 46.5	27.2	18 36.1	13.5	18 29.1	11.4	55.1
05	256 46.7	26.2	33 08.6	13.4	18 40.5	11.3	55.1
06	271 46.9	N 0 25.3	47 41.0	13.4	N18 51.8	11.3	55.0
07	286 47.1	24.3	62 13.4	13.3	19 03.1	11.1	55.0
08	301 47.3	23.3	76 45.7	13.3	19 14.2	11.1	55.0
09	316 47.6	· · 22.4	91 18.0	13.2	19 25.3	10.9	55.0
10	331 47.8	21.4	105 50.2	13.2	19 36.2	10.9	55.0
11	346 48.0	20.4	120 22.4	13.2	19 47.1	10.8	54.9
12	1 48.2	N 0 19.4	134 54.6	13.1	N19 57.9	10.7	54.9
13	16 48.4	18.5	149 26.7	13.1	20 08.6	10.6	54.9
14	31 48.7	17.5	163 58.8	13.0	20 19.2	10.5	54.9
15	46 48.9	· · 16.5	178 30.8	12.9	20 29.7	10.4	54.9
16	61 49.1	15.6	193 02.7	13.0	20 40.1	10.3	54.8
17	76 49.3	14.6	207 34.7	12.8	20 50.4	10.2	54.8
18	91 49.5	N 0 13.6	222 06.5	12.9	N21 00.6	10.1	54.8
19	106 49.8	12.6	236 38.4	12.7	21 10.7	10.0	54.8
20	121 50.0	11.7	251 10.1	12.8	21 20.7	09.9	54.8
21	136 50.2	· · 10.7	265 41.9	12.7	21 30.6	09.9	54.8
22	151 50.4	09.7	280 13.6	12.6	21 40.5	09.7	54.7
23	166 50.6	08.7	294 45.2	12.6	21 50.2	09.6	54.7
23 00 (TUESDAY)	181 50.9	N 0 07.8	309 16.8	12.5	N21 59.8	09.5	54.7
01	196 51.1	06.8	323 48.3	12.5	22 09.3	09.4	54.7
02	211 51.3	05.8	338 19.8	12.4	22 18.7	09.3	54.7
03	226 51.5	· · 04.8	352 51.2	12.4	22 28.0	09.2	54.7
04	241 51.7	03.9	7 22.6	12.4	22 37.2	09.1	54.7
05	256 52.0	02.9	21 54.0	12.3	22 46.3	08.9	54.6
06	271 52.2	N 0 01.9	36 25.3	12.2	N22 55.2	08.9	54.6
07	286 52.4	N 0 01.0	50 56.5	12.2	23 04.1	08.8	54.6
08	301 52.6	0 00.0	65 27.7	12.1	23 12.9	08.6	54.6
09	316 52.8	S 0 01.0	79 58.8	12.1	23 21.5	08.6	54.6
10	331 53.1	02.0	94 29.9	12.1	23 30.1	08.4	54.6
11	346 53.3	02.9	109 01.0	12.0	23 38.5	08.4	54.6
12	1 53.5	S 0 03.9	123 32.0	11.9	N23 46.9	08.2	54.5
13	16 53.7	04.9	138 02.9	11.9	23 55.1	08.1	54.5
14	31 53.9	05.9	152 33.8	11.8	24 03.2	08.0	54.5
15	46 54.2	· · 06.8	167 04.6	11.8	24 11.2	07.9	54.5
16	61 54.4	07.8	181 35.4	11.8	24 19.1	07.7	54.5
17	76 54.6	08.8	196 06.2	11.7	24 26.8	07.7	54.5
18	91 54.8	S 0 09.7	210 36.9	11.6	N24 34.5	07.5	54.5
19	106 55.0	10.7	225 07.5	11.6	24 42.0	07.4	54.5
20	121 55.3	11.7	239 38.1	11.6	24 49.4	07.4	54.4
21	136 55.5	· · 12.7	254 08.7	11.5	24 56.8	07.1	54.4
22	151 55.7	13.6	268 39.2	11.5	25 03.9	07.1	54.4
23	166 55.9	14.6	283 09.7	11.4	25 11.0	07.0	54.4
24 00 (WEDNESDAY)	181 56.1	S 0 15.6	297 40.1	11.3	N25 18.0	06.8	54.4
01	196 56.4	16.6	312 10.4	11.3	25 24.8	06.7	54.4
02	211 56.6	17.5	326 40.7	11.3	25 31.5	06.6	54.4
03	226 56.8	· · 18.5	341 11.0	11.2	25 38.1	06.5	54.4
04	241 57.0	19.5	355 41.2	11.2	25 44.6	06.4	54.4
05	256 57.2	20.5	10 11.4	11.1	25 51.0	06.2	54.4
06	271 57.4	S 0 21.4	24 41.5	11.1	N25 57.2	06.1	54.3
07	286 57.7	22.4	39 11.6	11.1	26 03.3	06.0	54.3
08	301 57.9	23.4	53 41.7	11.0	26 09.3	05.9	54.3
09	316 58.1	· · 24.3	68 11.7	10.9	26 15.2	05.7	54.3
10	331 58.3	25.3	82 41.6	10.9	26 20.9	05.6	54.3
11	346 58.5	26.3	97 11.5	10.9	26 26.5	05.5	54.3
12	1 58.8	S 0 27.3	111 41.4	10.8	N26 32.0	05.4	54.3
13	16 59.0	28.2	126 11.2	10.8	26 37.4	05.2	54.3
14	31 59.2	29.2	140 41.0	10.8	26 42.6	05.2	54.3
15	46 59.4	· · 30.2	155 10.8	10.7	26 47.8	05.0	54.3
16	61 59.6	31.2	169 40.5	10.6	26 52.8	04.8	54.3
17	76 59.8	32.1	184 10.1	10.7	26 57.6	04.8	54.3
18	92 00.1	S 0 33.1	198 39.8	10.6	N27 02.4	04.6	54.3
19	107 00.3	34.1	213 09.4	10.5	27 07.0	04.4	54.3
20	122 00.5	35.1	227 38.9	10.5	27 11.4	04.4	54.3
21	137 00.7	· · 36.0	242 08.4	10.5	27 15.8	04.2	54.3
22	152 00.9	37.0	256 37.9	10.4	27 20.0	04.1	54.3
23	167 01.1	38.0	271 07.3	10.5	27 24.1	04.0	54.2
	S.D. 16.0	d 1.0	S.D. 15.0		14.9		14.8

Lat. °	Twilight Naut. h m	Twilight Civil h m	Sunrise h m	Moonrise 22 h m	Moonrise 23 h m	Moonrise 24 h m	Moonrise 25 h m
N 72	03 02	04 33	05 41	□	□	□	□
N 70	03 22	04 41	05 42	□	□	□	□
68	03 37	04 47	05 43	16 33	□	□	□
66	03 49	04 53	05 44	17 16	□	□	□
64	03 59	04 57	05 45	17 45	17 23	□	□
62	04 07	05 01	05 45	18 08	18 02	17 54	□
60	04 14	05 04	05 46	18 26	18 29	18 39	19 04
N 58	04 20	05 07	05 46	18 41	18 51	19 09	19 41
56	04 25	05 09	05 46	18 54	19 09	19 32	20 08
54	04 29	05 11	05 47	19 05	19 24	19 50	20 29
52	04 33	05 13	05 47	19 16	19 37	20 06	20 47
50	04 37	05 15	05 47	19 25	19 48	20 20	21 02
45	04 44	05 18	05 48	19 44	20 12	20 48	21 32
N 40	04 49	05 21	05 48	19 59	20 31	21 10	21 56
35	04 54	05 23	05 48	20 13	20 48	21 28	22 15
30	04 57	05 25	05 49	20 24	21 02	21 44	22 32
20	05 01	05 27	05 49	20 44	21 26	22 11	23 00
N 10	05 04	05 28	05 49	21 02	21 46	22 34	23 24
0	05 05	05 29	05 49	21 18	22 06	22 56	23 47
S 10	05 04	05 28	05 49	21 35	22 26	23 18	24 10
20	05 01	05 27	05 49	21 53	22 47	23 41	24 34
30	04 57	05 25	05 49	22 13	23 11	24 08	00 08
35	04 54	05 23	05 49	22 25	23 26	24 24	00 24
40	04 50	05 21	05 48	22 39	23 43	24 43	00 43
45	04 44	05 19	05 48	22 56	24 03	00 03	01 06
S 50	04 37	05 15	05 48	23 17	24 29	00 29	01 36
52	04 34	05 14	05 47	23 27	24 41	00 41	01 50
54	04 30	05 12	05 47	23 38	24 55	00 55	02 07
56	04 26	05 10	05 47	23 50	25 12	01 12	02 27
58	04 21	05 07	05 46	24 05	00 05	01 32	02 53
S 60	04 15	05 05	05 46	24 23	00 23	01 58	03 28

Lat. °	Sunset h m	Twilight Civil h m	Twilight Naut. h m	Moonset 22 h m	Moonset 23 h m	Moonset 24 h m	Moonset 25 h m
N 72	18 01	19 09	20 38	□	□	□	□
N 70	18 00	19 01	20 19	□	□	□	□
68	18 00	18 55	20 05	13 23	□	□	□
66	17 59	18 50	19 53	12 41	□	□	□
64	17 59	18 46	19 44	12 12	14 13	□	□
62	17 58	18 42	19 36	11 51	13 35	15 26	□
60	17 58	18 39	19 29	11 34	13 08	14 41	16 02
N 58	17 58	18 37	19 23	11 19	12 47	14 11	15 24
56	17 57	18 34	19 18	11 07	12 30	13 49	14 58
54	17 57	18 32	19 14	10 56	12 15	13 30	14 37
52	17 57	18 31	19 10	10 47	12 03	13 15	14 19
50	17 57	18 29	19 07	10 38	11 51	13 01	14 04
45	17 56	18 26	19 00	10 20	11 28	12 34	13 34
N 40	17 56	18 23	18 55	10 06	11 10	12 12	13 11
35	17 56	18 21	18 51	09 54	10 55	11 54	12 51
30	17 56	18 20	18 47	09 43	10 41	11 39	12 35
20	17 56	18 18	18 43	09 25	10 19	11 13	12 07
N 10	17 56	18 17	18 41	09 09	09 59	10 51	11 43
0	17 56	18 16	18 40	08 54	09 41	10 30	11 20
S 10	17 56	18 17	18 41	08 40	09 23	10 09	10 58
20	17 56	18 18	18 44	08 24	09 04	09 47	10 34
30	17 57	18 20	18 48	08 06	08 41	09 21	10 06
35	17 57	18 22	18 52	07 56	08 28	09 06	09 50
40	17 57	18 24	18 56	07 44	08 13	08 49	09 31
45	17 58	18 27	19 01	07 30	07 56	08 28	09 07
S 50	17 58	18 31	19 09	07 13	07 34	08 01	08 38
52	17 59	18 32	19 12	07 05	07 24	07 48	08 23
54	17 59	18 34	19 16	06 57	07 12	07 34	08 06
56	17 59	18 36	19 21	06 47	06 59	07 17	07 45
58	18 00	18 39	19 26	06 36	06 43	06 56	07 20
S 60	18 00	18 42	19 32	06 23	06 25	06 30	06 44

Day	SUN Eqn. of Time 00ʰ m s	SUN Eqn. of Time 12ʰ m s	SUN Mer. Pass. h m	MOON Mer. Pass. Upper h m	MOON Mer. Pass. Lower h m	MOON Age d	MOON Phase
22	07 02	07 12	11 53	02 43	15 06	18	◑
23	07 23	07 34	11 52	03 30	15 53	19	
24	07 44	07 55	11 52	04 18	16 43	20	

FIGURE 9.4 ☆ *NAUTICAL ALMANAC*, RIGHT HAND DAILY PAGE (TYPICAL)

is moved to the Total space under Ha and rewritten: 1°04.2′. Add 1°04.2′ to Ha, round off to the nearest minute, obtaining and recording: Ho: 21°10′.

Go to the yellow pages in the almanac headed: 58m.

Go down the column headed: MOON corresponding to 38s, obtaining and recording: Min. & Sec.: 13°59.4′; little d correction: Corr±: 10.8′ circling the + sign, and v correction: Corr.: 13.0′.

Complete the worksheet in the usual manner and plot the resulting LOP. Label the LOP: 2200/MOON.

On a moonlit night, the moon sight could be combined with any number of star sights that could be brought to the visible portion of the horizon. As previously mentioned, however, the portion of the horizon within an azimuth of approximately ± 20° of to the moon should be avoided.

My favorite moon sight is when the moon is visible during daylight hours and can be combined with a sun sight.

About an hour before sighting the moon I had decided to do some preparation to determine if it would be possible to observe a planet.

Open the 1986 almanac to pages 8 and 9 or go to Figure 9.5 headed: PLANET NOTES, 1986, and Figure 9.6 headed: PLANETS, 1986.

Read the information contained in the PLANET NOTES, 1986 under VISIBILITY OF PLANETS.

After reading the information contained under: JUPITER, we decided to sight it. Jupiter would be at opposition (opposite the sun) and could be seen all night on September 10th. Jupiter would be visible for most of the night on September 21st.

By inspecting the graph PLANETS, 1986, we see that Jupiter's meridian passage would be at approximately 2300 hours local time. We would be sighting the planet about an hour before this, at approximately 2200 hours.

Another use of the Star Finder, #2102-D, is to locate the planets by altitude and azimuth as you have done for the stars.

The only problem is that we have to manually plot the positions of the planets on the base of the star finder. The plotting is necessary because the planets, unlike the stars, do not maintain their positions in the heavens with respect to the earth. The planets, like the earth, are rotating around the sun and describe a complex path when observed from the earth.

A plotted position of a planet, on the base, is useful for a few

8 PLANET NOTES, 1986

VISIBILITY OF PLANETS

VENUS is too close to the Sun for observation from the beginning of the year until early March, and can then be seen as a brilliant object in the evening sky until early November, when it again becomes too close to the Sun for observation. Before mid-November it reappears as a morning star and can be seen in the morning sky for the rest of the year. Venus is in conjunction with Mercury on March 8 and October 18.

MARS rises well after midnight at the beginning of the year in Libra. Its westward elongation gradually increases, and in February and March it moves through Scorpius, Ophiuchus (passing 5°N. of *Antares* on February 17) and into Sagittarius, and is at opposition on July 10 when it can be seen throughout the night. It remains in Sagittarius until early October and then moves through Capricornus, Aquarius and into Pisces in late December. From mid-November until the end of the year it can only be seen in the evening sky. Mars is in conjunction with Saturn on February 18 and with Jupiter on December 19.

JUPITER can be seen in the evening sky in Capricornus from the beginning of the year until early February, when it becomes too close to the Sun for observation. It reappears in the morning sky early in March in Aquarius, in which constellation it remains for the rest of the year. Jupiter is at opposition on September 10 when it can be seen throughout the night, and from early December until the end of the year it can only be seen in the evening sky. Jupiter is in conjunction with Mars on December 19.

SATURN rises before sunrise at the beginning of the year in Scorpius and moves into Ophiuchus in mid-January, (passing 7°N. of *Antares* on February 10 and again on April 26). It returns to Scorpius in late May, is at opposition on May 28 when it can be seen throughout the night, and in mid-October returns to Ophiuchus, in which constellation it remains for the rest of the year, (passing 6°N. of *Antares* on November 3). From late August until shortly after mid-November it can only be seen in the evening sky and then becomes too close to the Sun for observation until late December, when it reappears in the morning sky. Saturn is in conjunction with Mars on February 18.

MERCURY can only be seen low in the east before sunrise, or low in the west after sunset (about the time of beginning or end of civil twilight). It is visible in the mornings between the following approximate dates: January 1 (−0·3) to January 17 (−0·5); March 24 (+1·9) to May 15 (−1·2); July 31 (+2·0) to August 28 (−1·3); November 19 (+1·2) to December 27 (−0·5); the planet is brighter at the end of each period. It is visible in the evenings between the following approximate dates: February 13 (−1·1) to March 10 (+1·6); May 31 (−1·3) to July 16 (+2·2); September 17 (−0·7) to November 7 (+1·3); the planet is brighter at the beginning of each period. The figures in parentheses are the magnitudes. Mercury transits the Sun's disk on November 13 from 01^h 43^m to 06^h 31^m; the event is visible from the Pacific Ocean except the eastern part, Australasia, Asia, the Indian Ocean, part of Antarctica, Africa except the north-western part and eastern Europe.

PLANET DIAGRAM

General Description. The diagram on the opposite page shows, in graphical form for any date during the year, the local mean time of meridian passage of the Sun, of the five planets Mercury, Venus, Mars, Jupiter, and Saturn, and of each 30° of S.H.A.; intermediate lines, corresponding to particular stars, may be drawn in by the user if he so desires. It is intended to provide a general picture of the availability of planets and stars for observation.

On each side of the line marking the time of meridian passage of the Sun a band, 45^m wide, is shaded to indicate that planets and most stars crossing the meridian within 45^m of the Sun are too close to the Sun for observation.

Method of use and interpretation. For any date the diagram provides immediately the local mean times of meridian passage of the Sun, planets and stars, and thus the following information:

(a) whether a planet or star is too close to the Sun for observation;

(b) some indication of its position in the sky, especially during twilight;

(c) the proximity of other planets.

When the meridian passage of an outer planet occurs at midnight the body is in opposition to the Sun and is visible all night; a planet may then be observable during both morning and evening twilights. As the time of meridian passage decreases, the body eventually ceases to be observable in the morning, but its altitude above the eastern horizon at sunset gradually increases; this continues until the body is on the meridian during evening twilight. From then onwards the body is observable above the western horizon and its altitude at sunset gradually decreases; eventually the body becomes too close to the Sun for observation. When the body again becomes visible it is seen low in the east during morning twilight; its altitude at sunrise increases until meridian passage occurs during morning twilight. Then, as the time of meridian passage decreases to 0^h, the body is observable in the west during morning twilight with a gradually decreasing altitude, until it once again reaches opposition.

DO NOT CONFUSE

Mars with Saturn in February and with Jupiter in December; on both occasions Mars is the fainter object. The reddish tint of Mars should assist in its identification.

Venus with Mercury in early March and late October; on both occasions Venus is the brighter object.

Mercury with Saturn in late December when Mercury is the brighter object.

FIGURE 9.5 ☆ *NAUTICAL ALMANAC*, PLANET NOTES (TYPICAL)

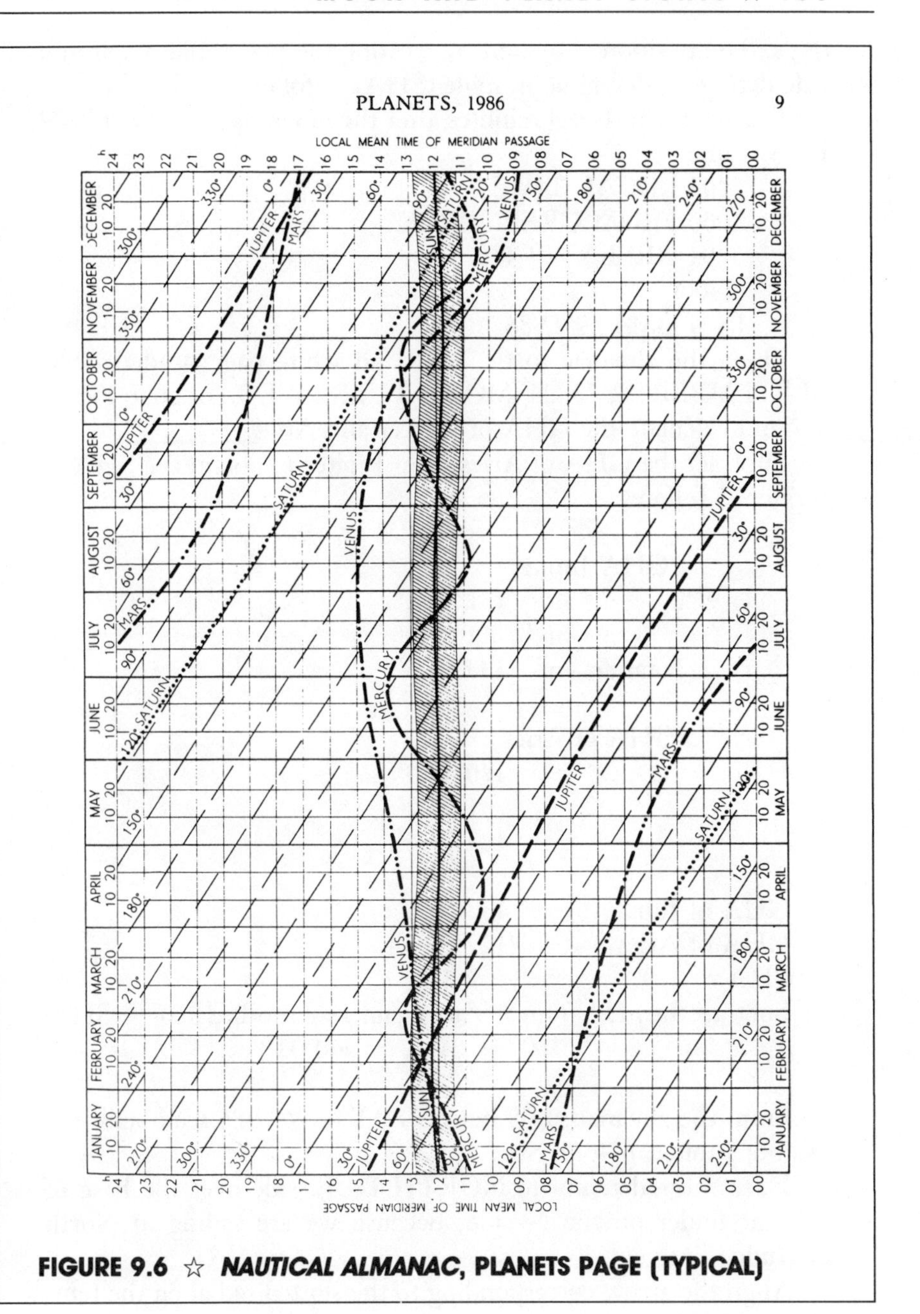

FIGURE 9.6 ☆ *NAUTICAL ALMANAC*, PLANETS PAGE (TYPICAL)

days. The position is plotted by treating the planet like a star and calculating a sidereal hour angle (SHA) as follows:

Calculate GMT at 4 minutes after the moon sight at about 2204 Local Time:

2204 LOCAL TIME
1000 FROM GREENWICH.
3204 GREENWICH.
(2400)
0804 GMT, 9/22/86.

Open the almanac to the left hand white page headed: 1986 SEPTEMBER 22, 23, 24 (MON., TUES., WED.) or go to Figure 9.7 and calculate the GHA of Jupiter and Aries.

Subtract the GHA of Aries from Jupiter to obtain the SHA of Jupiter as follows:

GHA Jupiter

9/22	8h	132°50.6′
	4m	1°00.0′
Total 8h 4m		133°50.6′

GHA of Aries

120°53.4′
1°00.2′
123°53.6′

GHA of Jup.	133°50.6′
GHA of Aries	(123°53.6′)
SHA	9°57′

Subtract the SHA from 360° to obtain a value called 360° − SHA:

360° − SHA = 350°03′

From the almanac at 08 hours GMT on 9/22/86 we obtain the Dec. of Jupiter: 6°46.6′ S.

Place the red template NORTH LAT. side over the base of the star finder on the N side, because we are sailing in North Latitude.

Align the arrow corresponding to the slotted radial on the template with 350°.

If the declination were the same name as the latitude we would plot toward the center where the lines are solid.

Because the declination is contrary to latitude the declination

186 1986 SEPTEMBER 22, 23, 24 (MON., TUES., WED.)

	G.M.T. (UT)	ARIES	VENUS −4.2		MARS −0.9		JUPITER −2.4		SATURN +0.8		STARS		
		G.H.A.	G.H.A.	Dec.	G.H.A.	Dec.	G.H.A.	Dec.	G.H.A.	Dec.	Name	S.H.A.	Dec.
	d h	° '	° '	° '	° '	° '	° '	° '	° '	° '		° '	° '
	22 00	0 33.7	143 15.2	S19 56.1	66 39.1	S25 34.0	12 28.5	S 6 45.6	117 29.6	S19 27.4	Acamar	315 33.3	S40 21.1
	01	15 36.2	158 16.0	56.9	81 40.4	33.7	27 31.3	45.7	132 31.9	27.4	Achernar	335 41.0	S57 18.0
	02	30 38.6	173 16.9	57.7	96 41.6	33.4	42 34.0	45.8	147 34.2	27.4	Acrux	173 33.2	S63 01.5
	03	45 41.1	188 17.8	· · 58.5	111 42.9	· · 33.1	57 36.8	· · 45.9	162 36.4	· · 27.5	Adhara	255 28.6	S28 56.8
	04	60 43.6	203 18.7	19 59.3	126 44.2	32.8	72 39.6	46.1	177 38.7	27.5	Aldebaran	291 12.7	N16 29.2
	05	75 46.0	218 19.5	20 00.1	141 45.4	32.5	87 42.3	46.2	192 41.0	27.6			
	06	90 48.5	233 20.4	S20 00.9	156 46.7	S25 32.2	102 45.1	S 6 46.3	207 43.3	S19 27.6	Alioth	166 38.7	N56 02.0
	07	105 51.0	248 21.3	01.7	171 48.0	32.0	117 47.8	46.4	222 45.6	27.6	Alkaid	153 15.1	N49 22.9
	08	120 53.4	263 22.2	02.5	186 49.2	31.7	132 50.6	46.6	237 47.9	27.7	Al Na'ir	28 08.6	S47 01.7
M	09	135 55.9	278 23.1	· · 03.2	201 50.5	· · 31.4	147 53.3	· · 46.7	252 50.1	· · 27.7	Alnilam	276 07.0	S 1 12.3
O	10	150 58.3	293 24.0	04.0	216 51.8	31.1	162 56.1	46.8	267 52.4	27.8	Alphard	218 16.3	S 8 35.8
N	11	166 00.8	308 24.8	04.8	231 53.0	30.8	177 58.9	46.9	282 54.7	27.8			
D	12	181 03.3	323 25.7	S20 05.6	246 54.3	S25 30.5	193 01.6	S 6 47.0	297 57.0	S19 27.8	Alphecca	126 28.4	N26 45.7
A	13	196 05.7	338 26.6	06.4	261 55.5	30.2	208 04.4	47.2	312 59.3	27.9	Alpheratz	358 04.3	N29 01.1
Y	14	211 08.2	353 27.5	07.1	276 56.8	29.9	223 07.1	47.3	328 01.6	27.9	Altair	62 27.9	N 8 50.0
	15	226 10.7	8 28.4	· · 07.9	291 58.1	· · 29.6	238 09.9	· · 47.4	343 03.8	· · 27.9	Ankaa	353 35.1	S42 22.6
	16	241 13.1	23 29.3	08.7	306 59.3	29.3	253 12.6	47.5	358 06.1	28.0	Antares	112 51.4	S26 24.3
	17	256 15.6	38 30.2	09.5	322 00.6	29.0	268 15.4	47.6	13 08.4	28.0			
	18	271 18.1	53 31.1	S20 10.3	337 01.8	S25 28.7	283 18.1	S 6 47.8	28 10.7	S19 28.1	Arcturus	146 14.5	N19 15.2
	19	286 20.5	68 32.0	11.0	352 03.1	28.4	298 20.9	47.9	43 13.0	28.1	Atria	108 11.8	S69 00.6
	20	301 23.0	83 32.9	11.8	7 04.4	28.1	313 23.7	48.0	58 15.3	28.1	Avior	234 26.9	S59 27.6
	21	316 25.5	98 33.8	· · 12.6	22 05.6	· · 27.8	328 26.4	· · 48.1	73 17.5	· · 28.2	Bellatrix	278 53.8	N 6 20.5
	22	331 27.9	113 34.7	13.4	37 06.9	27.5	343 29.2	48.3	88 19.8	28.2	Betelgeuse	271 23.3	N 7 24.6
	23	346 30.4	128 35.6	14.1	52 08.1	27.2	358 31.9	48.4	103 22.1	28.3			
	23 00	1 32.8	143 36.5	S20 14.9	67 09.4	S25 26.9	13 34.7	S 6 48.5	118 24.4	S19 28.3	Canopus	264 05.3	S52 40.8
	01	16 35.3	158 37.4	15.7	82 10.6	26.6	28 37.4	48.6	133 26.7	28.3	Capella	281 04.5	N45 59.1
	02	31 37.8	173 38.4	16.4	97 11.9	26.3	43 40.2	48.7	148 29.0	28.4	Deneb	49 45.2	N45 14.1
	03	46 40.2	188 39.3	· · 17.2	112 13.1	· · 26.0	58 42.9	· · 48.9	163 31.2	· · 28.4	Denebola	182 54.6	N14 39.0
	04	61 42.7	203 40.2	18.0	127 14.4	25.7	73 45.7	49.0	178 33.5	28.5	Diphda	349 15.9	S18 03.4
	05	76 45.2	218 41.1	18.7	142 15.6	25.4	88 48.5	49.1	193 35.8	28.5			
	06	91 47.6	233 42.0	S20 19.5	157 16.9	S25 25.2	103 51.2	S 6 49.2	208 38.1	S19 28.5	Dubhe	194 16.8	N61 49.4
	07	106 50.1	248 43.0	20.3	172 18.1	24.9	118 54.0	49.3	223 40.4	28.6	Elnath	278 38.3	N28 35.9
T	08	121 52.6	263 43.9	21.0	187 19.4	24.6	133 56.7	49.5	238 42.6	28.6	Eltanin	90 55.6	N51 29.6
U	09	136 55.0	278 44.8	· · 21.8	202 20.6	· · 24.3	148 59.5	· · 49.6	253 44.9	· · 28.7	Enif	34 06.9	N 9 48.9
E	10	151 57.5	293 45.7	22.6	217 21.9	24.0	164 02.2	49.7	268 47.2	28.7	Fomalhaut	15 45.9	S29 41.6
S	11	166 59.9	308 46.7	23.3	232 23.1	23.7	179 05.0	49.8	283 49.5	28.7			
D	12	182 02.4	323 47.6	S20 24.1	247 24.3	S25 23.4	194 07.7	S 6 49.9	298 51.8	S19 28.8	Gacrux	172 24.6	S57 02.3
A	13	197 04.9	338 48.5	24.9	262 25.6	23.1	209 10.5	50.1	313 54.1	28.8	Gienah	176 13.7	S17 27.9
Y	14	212 07.3	353 49.5	25.6	277 26.8	22.8	224 13.2	50.2	328 56.3	28.9	Hadar	149 17.7	S60 18.7
	15	227 09.8	8 50.4	· · 26.4	292 28.1	· · 22.5	239 16.0	· · 50.3	343 58.6	· · 28.9	Hamal	328 23.5	N23 24.1
	16	242 12.3	23 51.3	27.1	307 29.3	22.2	254 18.8	50.4	359 00.9	28.9	Kaus Aust.	84 10.7	S34 23.7
	17	257 14.7	38 52.3	27.9	322 30.6	21.9	269 21.5	50.5	14 03.2	29.0			
	18	272 17.2	53 53.2	S20 28.6	337 31.8	S25 21.5	284 24.3	S 6 50.7	29 05.5	S19 29.0	Kochab	137 19.5	N74 12.8
	19	287 19.7	68 54.1	29.4	352 33.0	21.2	299 27.0	50.8	44 07.7	29.1	Markab	13 58.4	N15 08.1
	20	302 22.1	83 55.1	30.1	7 34.3	20.9	314 29.8	50.9	59 10.0	29.1	Menkar	314 36.1	N 4 02.5
	21	317 24.6	98 56.0	· · 30.9	22 35.5	· · 20.6	329 32.5	· · 51.0	74 12.3	· · 29.1	Menkent	148 32.1	S36 18.3
	22	332 27.1	113 57.0	31.7	37 36.8	20.3	344 35.3	51.1	89 14.6	29.2	Miaplacidus	221 45.1	S69 39.4
	23	347 29.5	128 57.9	32.4	52 38.0	20.0	359 38.0	51.3	104 16.8	29.2			
	24 00	2 32.0	143 58.9	S20 33.2	67 39.2	S25 19.7	14 40.8	S 6 51.4	119 19.1	S19 29.3	Mirfak	309 09.4	N49 48.9
	01	17 34.4	158 59.8	33.9	82 40.5	19.4	29 43.5	51.5	134 21.4	29.3	Nunki	76 23.4	S26 19.0
	02	32 36.9	174 00.8	34.7	97 41.7	19.1	44 46.3	51.6	149 23.7	29.3	Peacock	53 50.6	S56 47.0
	03	47 39.4	189 01.7	· · 35.4	112 42.9	· · 18.8	59 49.0	· · 51.7	164 26.0	· · 29.4	Pollux	243 52.7	N28 03.7
	04	62 41.8	204 02.7	36.1	127 44.2	18.5	74 51.8	51.9	179 28.2	29.4	Procyon	245 21.1	N 5 15.8
	05	77 44.3	219 03.7	36.9	142 45.4	18.2	89 54.5	52.0	194 30.5	29.5			
	06	92 46.8	234 04.6	S20 37.6	157 46.6	S25 17.9	104 57.3	S 6 52.1	209 32.8	S19 29.5	Rasalhague	96 25.4	N12 34.2
W	07	107 49.2	249 05.6	38.4	172 47.9	17.6	120 00.0	52.2	224 35.1	29.5	Regulus	208 05.4	N12 02.2
E	08	122 51.7	264 06.6	39.1	187 49.1	17.3	135 02.8	52.3	239 37.4	29.6	Rigel	281 31.5	S 8 12.7
D	09	137 54.2	279 07.5	· · 39.9	202 50.3	· · 17.0	150 05.5	· · 52.4	254 39.6	· · 29.6	Rigil Kent.	140 20.4	S60 47.0
N	10	152 56.6	294 08.5	40.6	217 51.6	16.7	165 08.3	52.6	269 41.9	29.7	Sabik	102 36.0	S15 42.6
E	11	167 59.1	309 09.5	41.3	232 52.8	16.4	180 11.1	52.7	284 44.2	29.7			
S	12	183 01.6	324 10.4	S20 42.1	247 54.0	S25 16.1	195 13.8	S 6 52.8	299 46.5	S19 29.8	Schedar	350 03.6	N56 27.9
D	13	198 04.0	339 11.4	42.8	262 55.2	15.8	210 16.6	52.9	314 48.7	29.8	Shaula	96 49.6	S37 05.9
A	14	213 06.5	354 12.4	43.6	277 56.5	15.5	225 19.3	53.0	329 51.0	29.8	Sirius	258 51.7	S16 41.5
Y	15	228 08.9	9 13.4	· · 44.3	292 57.7	· · 15.2	240 22.1	· · 53.2	344 53.3	· · 29.9	Spica	158 53.1	S11 05.4
	16	243 11.4	24 14.3	45.0	307 58.9	14.9	255 24.8	53.3	359 55.6	29.9	Suhail	223 07.9	S43 22.3
	17	258 13.9	39 15.3	45.8	323 00.1	14.5	270 27.6	53.4	14 57.8	30.0			
	18	273 16.3	54 16.3	S20 46.5	338 01.4	S25 14.2	285 30.3	S 6 53.5	30 00.1	S19 30.0	Vega	80 52.7	N38 46.4
	19	288 18.8	69 17.3	47.2	353 02.6	13.9	300 33.1	53.6	45 02.4	30.0	Zuben'ubi	137 28.3	S15 59.2
	20	303 21.3	84 18.3	48.0	8 03.8	13.6	315 35.8	53.7	60 04.7	30.1		S.H.A.	Mer. Pass.
	21	318 23.7	99 19.3	· · 48.7	23 05.0	· · 13.3	330 38.6	· · 53.9	75 07.0	· · 30.1		° '	h m
	22	333 26.2	114 20.3	49.4	38 06.3	13.0	345 41.3	54.0	90 09.2	30.2	Venus	142 03.7	14 25
	23	348 28.7	129 21.3	50.2	53 07.5	12.7	0 44.1	54.1	105 11.5	30.2	Mars	65 36.5	19 30
	Mer. Pass.	h m 23 49.9	v 0.9	d 0.8	v 1.2	d 0.3	v 2.8	d 0.1	v 2.3	d 0.0	Jupiter	12 01.8	23 01
											Saturn	116 51.6	16 04

FIGURE 9.7 ☆ *NAUTICAL ALMANAC*, LEFT HAND DAILY PAGE (TYPICAL)

is plotted outward in the dotted lines. Going outward from 0°, with a pencil, make a dot at approximately 7° in the slot and next to the radial with the arrow. Because the declination is in small graduations of 2° it is not possible to plot with greater accuracy.

Remove the template, draw a small circle around the dot, and label it Jupiter.

With Jupiter plotted on the star finder, you will be able to determine the approximate altitude and azimuth for Jupiter in the same manner as the stars.

Calculate the LHA of Aries at 0804 GMT on 9/22/86 as follows:

9/22	8h	120°53.4′
	4m	1°00.2′
		360°
GHA of Aries		481°53.6′
Ass. Long.		(150°53.6′)
LHA of Aries		331°

Place the template for the nearest latitude over the N side of the base. The closest template to the latitude is: LATITUDE 25°N.

Align the arrow on the template with 331° on the base, and record the altitude and azimuth of Jupiter.

I obtained: Hc: 53° and Zn: 148°. The approximate altitude and azimuth thus found is used for presetting the sextant and facing toward the planet.

Begin a worksheet and record as follows: No.: 45; Area: NORTH PACIFIC; Body: Jupiter; Date: 9/21/86; D.R. Lat.: 23°42′ N; and D.R. Long.: 151°04′ W as shown in Figure 9.8.

A stopwatch was started at 8h 2m, the sextant was checked for index error, and the mark corresponding to the micrometer drum on the index arm read: 58′. The sextant was preset at 53°.

I faced Jupiter by standing over the binnacle and observing the planet at the approximate azimuth while taking the earth's magnetic variation into consideration. Actually Jupiter was bright and stood out among the stars in the background.

I moved to the stern rail, attached myself to the rail via my safety harness; while viewing Jupiter through the telescope of the sextant above the moonlit horizon I adjusted the micrometer drum. When Jupiter touched the horizon we timed the sight and recorded: Stop Watch: 1m 55s and Hs: 55°09′.

My eye was approximately 11 feet above the water.

NO. 45 **LINE OF POSITION WORKSHEET** AREA NORTH PACIFIC

BODY	DATE	D. R. LAT.	D. R. LONG.	LOG	WWV	STOP WATCH	LOCAL TIME
JUPITER	9-21-86	N 23° 42'	W 151° 04'	—	8h 2m	1m 55s	2204

TIME	HR.	MIN.	SEC.
WWV	8	2	—
Stop Watch	—	1	55
GMT	8	3	55

ASS. POSITION LHA (ALMANAC)

GHA Hours	132° 50.6'		
Min. & Sec.	0° 58.8'		
Stars Only SHA	—		
Moon & Planets Only V:	2.8	Corr.	0.2'
360°	360°		
Total GHA	493° 50'		
Ass. Long.	150° 50'		
LHA	343°		

DECLINATION (ALMANAC)

Dec. Hours	S 6° 46.6'		
"d" ⊕	0.1'	Corr. ⊕	0.0'
True Dec.	S 6° 47'		

SUN, STARS, & PLANETS SEXTANT (ALMANAC)

Hs	55° 07'	Ha	55° 07.8'
IC	2'	Corr.	− 0.7'
Dip	− 3.2'	Add'l Corr.	—
Ha	55° 07.8'	Ho	55° 07'

INTERCEPT AZIMUTH – FIRST ENTRY

Lat. 24	LHA 343	A 15° 29'	A° 15	A' 29
		B ⊕ 65° 02'	Z_1 ⊕ 82.9°	
		Dec ⊖ 6° 47'		
	Sum: B+Dec.	F 58° 15'	F° 58	F' 15

INTERCEPT AZIMUTH – SECOND ENTRY

A° 15	F° 58	H 55° 00'	P° 63	Z_2 ⊕ 67.5°
F' 15	P° 63	$Corr._1$ ⊕ 13'	$Z_2°$ 68	
	Sum	55° 13'	Z_1 82.9°	
A' 29	$Z_2°$ 68	$Corr._2$ ⊖ 11'	Z_2 67.5°	
	True Hc	55° 02'	Z 150°	
	True Ho	55° 07'	Zn 150°	
	Intercept	05'	T ✓	A

MOON ONLY SEXTANT (ALMANAC)

Corr.			Hs
Add'l Corr.			IC
HP	Corr.		Dip
Total			Ha
		Total	
		Ho	

PLOTTING

Ass. Long.	W 150° 50'		
Zn 150°	Ass. Lat. N 24°		
Intercept	T 5'	A	

FIGURE 9.8 ☆

The sight reduction work is very simple for Jupiter. The altitude corrections are obtained from the STARS AND PLANETS column on the inside front cover or the white insert card in the almanac.

The column headed: JUPITER on the left hand white pages is used to obtain GHA and Dec.

Little v and d are obtained from the bottom of the column headed: JUPITER and their corrections are obtained from the INCREMENTS AND CORRECTIONS, i.e., the yellow pages for minutes and seconds. Be sure to use the column headed: SUN/PLANETS for minutes and seconds corrections.

Complete the sight reduction calculations. They are the same as for the sun, moon, and stars.

Plot the LOP, label the LOP: 2200/JUPITER.

Label the intersection of the moon and Jupiter LOPs horizontally: 2200 FIX and begin a DR course line bearing 240°.

Open your current year almanac to the left hand white pages for any date. Note that after the heading: ARIES there are headings:

VENUS, MARS, JUPITER, and SATURN. The tabulated information for the three additional planets is treated the same as Jupiter.

There may be additional altitude corrections for some of the planets during the year. The additional corrections are found in the second column of the column headed: STARS AND PLANETS located inside the front cover or on the white insert card furnished with the almanac and would be applied to the apparent altitude of the sight.

Mars, Jupiter, and Saturn are *superior planets*, i.e., they rotate around the sun outside of the earth's orbit around the sun, and Venus is an *inferior planet*, i.e., it rotates around the sun inside of the earth's orbit around the sun.

Most of the information in the almanac has been covered. Exceptions would be calendars, *eclipses*, standard times at different ports around the world, etc. Read the additional information at your leisure.

PILOT WHALES, NORTH PACIFIC

TEN

THE STAR FIX II

Toward morning on September 22, 1986, we sailed across the Tropic of Cancer and entered the tropics. The weather was beautiful. We were blessed with a fair trade wind and the sight of puffy little cumulus clouds moving across the sky.

At about 1000 we encountered an enormous pod of pilot whales. What a fine greeting! There were pilot whales all around us, as far as we could see. These creatures are often mistaken for dolphins. However, they are much larger than any dolphins we have seen. The whales are dark charcoal grey or black in color, and their foreheads are very prominent and rounded. The pilot whales appeared to be curious about the boat, and swam very close without ever touching her. For that I was grateful. These creatures were anywhere from 15 to 25 feet long and massive in appearance. The whales stayed with us for about an hour, then disappeared in a matter of seconds.

Shortly after this little ripple in nautical history, we began to navigate. We started by advancing our DR course line, taking a course line sight at 1109, and an LAN sight at 1208 which gave us a running fix. See Figure 9.3 for details.

We followed the same course for the remainder of the day and the following day until LAN. Then, at 1216 we obtained a running fix and changed the course to 235° true in anticipation of strong northwest currents while approaching Maui, our planned landfall. We were about 150 nautical miles from Maui, and determined that an intensification of our navigation would be appropriate.

As mentioned earlier, the log was inoperative, and I was forced to estimate our position by experience, which did not bother me. The breakdown of the log is just another example of why it is not possible to rely on mechanical, electrical, and electronic devices at sea.

The trade winds were increasing as we closed on the island. Other than puffy little clouds, the sky was clear near sunset, and I decided to observe a round of stars at twilight to obtain our position. I hoped to observe three stars, and labelled the worksheets as follows: No.: 50; 51; & 52 and Area: MAUI as shown in Figures 10.1, 10.2 and 10.3.

Begin with the first worksheet record: No.: 50; Date: 9/23/86.

Go to the lower right hand corner and record as follows: Write: SUNSET and open the almanac to the right hand white page headed: 1986 SEPTEMBER 22, 23, 24 (MON., TUES., WED.) or go to Figure 9.4.

Go to the lower right hand block headed: Sunset.

Go down the Lat. column to 20° N (nearest to our latitude), go across and record: 1843 TW—NAUT, 1818 TW—CIVIL, and subtract to obtain the working period: 25 minutes.

I did not bother with interpolating because the difference between 30° N and 20° N was only 2 minutes for Twilight Civil and 4 minutes for Twilight Nautical. One fifth of these numbers is less than 1 minute.

Divide the working period by 2, neglect fractions, and add 12 minutes to 1818 to obtain 1830 LOCAL TIME @ 150°, the middle of our working time at the full-hour meridian.

We knew that we had passed the 154° meridian shortly after LAN and would be about 30 miles beyond the 154° meridian at twilight. See Figure 10.4.

NO. 50 **LINE OF POSITION WORKSHEET** AREA MAUI

BODY	DATE	D. R. LAT.	D. R. LONG.	LOG	WWV	STOP WATCH	LOCAL TIME
DENEB	9-23-86	N 22° 00'	W 154° 22'	—			1844

TIME	HR.	MIN.	SEC.
WWV			
Stop Watch			
GMT	4	44	—

ASS. POSITION LHA (ALMANAC)

GHA Hours	♈ 62° 41.8'
Min. & Sec.	♈ 11° 01.8'
Stars Only SHA	—
Moon & Planets Only V: / Corr.	—
360°	360°
Total GHA	433° 44'
Ass. Long.	154° 44'
LHA	♈ 279°

DECLINATION (ALMANAC)

Dec. Hours	
"d" ±	Corr. ±
True Dec.	

SUN, STARS, & PLANETS SEXTANT (ALMANAC)

Hs	55° 46'	Ha	55° 44.8'
IC	2'	Corr.	-0.7'
Dip	-3.2'	Add'l Corr.	—
Ha	55° 44.8'	Ho	55° 44'

INTERCEPT AZIMUTH - FIRST ENTRY

Lat.	LHA	A	A°	A'
		B ±	Z_1 ±	
		Dec. ±		
Sum: B+Dec.		F	F°	F'

INTERCEPT AZIMUTH - SECOND ENTRY

A°	F°	H	P°	Z_2 ±
F'	P°	$Corr._1$ ±	Z_2°	
	Sum		Z_1	
A'	Z_2°	$Corr._2$ ±	Z_2	
	True Hc	55° 31'	Z	
	True Ho	55° 44'	Zn	
	Intercept	13'	T ✓	A

MOON ONLY SEXTANT (ALMANAC)

Corr.		Hs
Add'l Corr.		IC
HP	Corr.	Dip
Total		Ha
	Total	
	Ho	

PLOTTING

Ass. Long.	W 154° 44'		
Zn	40°	Ass. Lat.	N 22°
Intercept	T 13'	A	

SUNSET
1843 TW - NAUT.
(1818) TW - CIVIL
2 | 25
12
1818
1830 LOCAL TIME @ 150°
18 PLUS 4° 30'
1848 LOCAL TIME @ 154° 30'
1000 FROM GREENWICH
2848 SAME DAY
(2400) GREENWICH
0448 GMT, 9-24-86

FIGURE 10.1 ☆

Reconstruct the 1216 running fix on a new plotting sheet and advance the DR course line 30 nautical miles on a bearing of 235° true.

Determine and record: D. R. Long.—154°22 W and D. R. Lat.—22°00′ N on worksheet No. 50. This DR position represented my best estimate without a log.

The tables headed: CONVERSION OF ARC TO TIME will be used to convert the additional longitude to time. From the DR plot I figured 4.5° or 4°30° would be a close enough approximation to determine the additional time.

Go down the first column to 4°. The 4° converts to 16 minutes.

Go to the last table on the right half of the page.

Go down the minutes column to 30′. The 30′ converts to 2 minutes. The 2 minutes is added to 16 minutes and the total is 18 minutes. Add to obtain and record: 1848 LOCAL TIME @ 154°30′.

Add 10 hours FROM GREENWICH to arrive at 2848 @ GREENWICH.

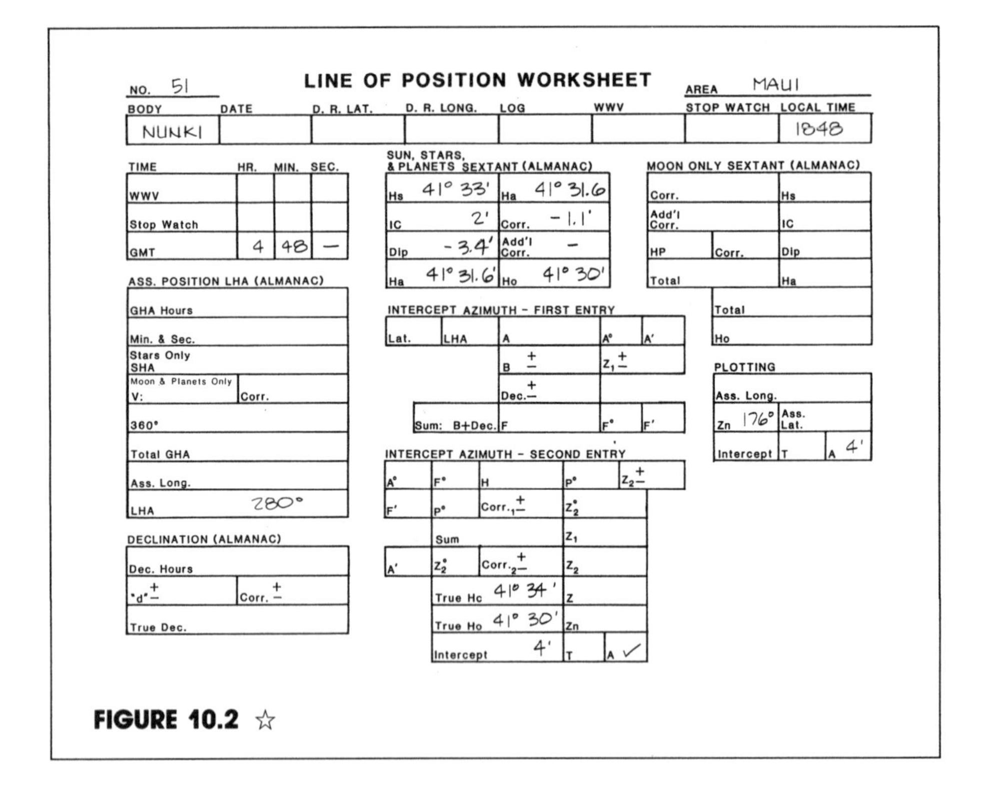

NO. 51 LINE OF POSITION WORKSHEET AREA MAUI

BODY	DATE	D. R. LAT.	D. R. LONG.	LOG	WWV	STOP WATCH	LOCAL TIME
NUNKI							1848

TIME	HR.	MIN.	SEC.
WWV			
Stop Watch			
GMT	4	48	—

SUN, STARS, & PLANETS SEXTANT (ALMANAC)

Hs	41° 33'	Ha	41° 31.6
IC	2'	Corr.	- 1.1'
Dip	- 3.4'	Add'l Corr.	—
Ha	41° 31.6'	Ho	41° 30'

MOON ONLY SEXTANT (ALMANAC)

Corr.		Hs	
Add'l Corr.		IC	
HP	Corr.	Dip	
Total		Ha	
	Total		
	Ho		

ASS. POSITION LHA (ALMANAC)

GHA Hours	
Min. & Sec.	
Stars Only SHA	
Moon & Planets Only V:	Corr.
360°	
Total GHA	
Ass. Long.	
LHA	280°

INTERCEPT AZIMUTH – FIRST ENTRY

Lat.	LHA	A	A°	A'
		B ±	Z_1 ±	
		Dec. ±		
	Sum: B+Dec.	F	F°	F'

PLOTTING

Ass. Long.			
Zn 176°	Ass. Lat.		
Intercept	T	A 4'	

INTERCEPT AZIMUTH – SECOND ENTRY

A°	F°	H	P°	Z_2 ±
F'	P°	$Corr._1$ ±	$Z_2°$	
	Sum		Z_1	
A'	$Z_2°$	$Corr._2$ ±	Z_2	
	True Hc 41° 34'		Z	
	True Ho 41° 30'		Zn	
	Intercept 4'		T	A ✓

DECLINATION (ALMANAC)

Dec. Hours	
"d" ±	Corr. ±
True Dec.	

FIGURE 10.2 ☆

Subtract 24 hours and obtain 0448 GMT 9/24/86, the time of the middle sight.

Record in the Time blocks on the three worksheets, in the spaces provided for GMT: 4 Hr. 44 Min.; 4 Hr. 48 Min.; and 4 Hr. 52 Min.

Record: Local Times: 1844, 1848, and 1852.

On worksheets No. 50, using GMT of 4h 44m, calculate the LHA of Aries. I obtained 279°.

On worksheets Nos. 51 and 52 record the LHA of Aries as 280° and 281° respectively.

In the Plotting block on No. 50 record the Ass. Long.: 154°44′ W and the Ass. Lat.: 22° N.

Several steps have already been saved. For example, you will not have to laboriously calculate an LHA for each star and you will be plotting all three stars from the same assumed position.

What stars, you may ask? Hold on to your hat! The best part is yet to come!

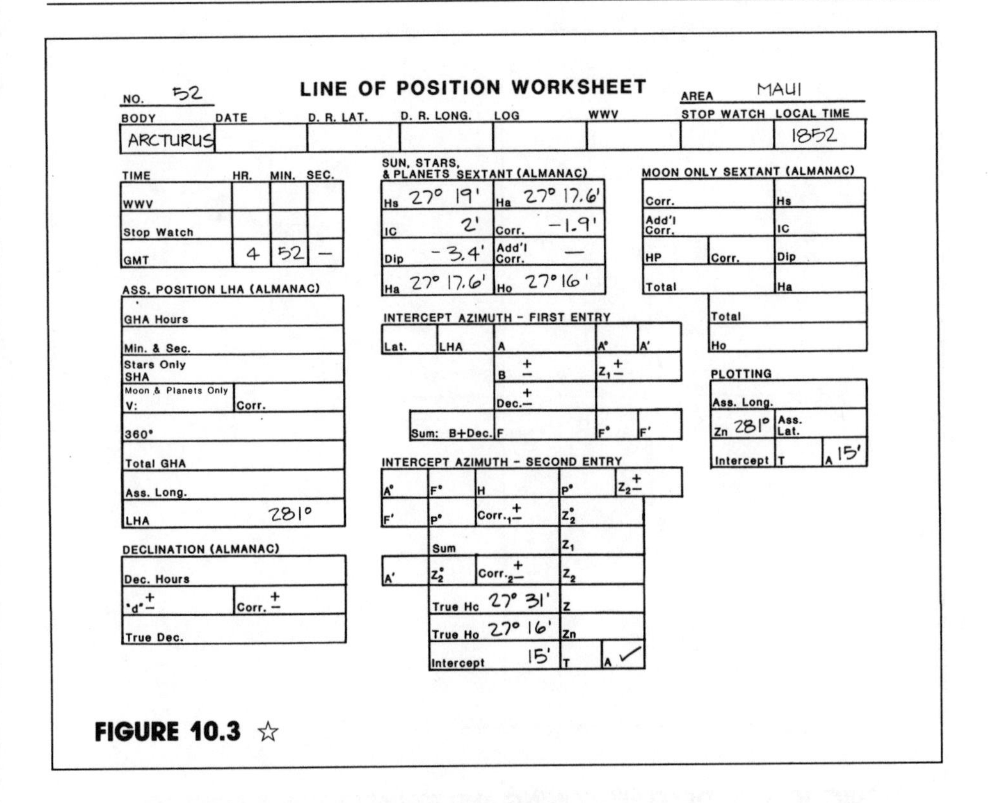

NO. 52 — LINE OF POSITION WORKSHEET — AREA MAUI

BODY	DATE	D. R. LAT.	D. R. LONG.	LOG	WWV	STOP WATCH	LOCAL TIME
ARCTURUS							1852

TIME	HR.	MIN.	SEC.
WWV			
Stop Watch			
GMT	4	52	—

SUN, STARS, & PLANETS SEXTANT (ALMANAC)

Hs	27° 19'	Ha	27° 17.6'
IC	2'	Corr.	−1.9'
Dip	−3.4'	Add'l Corr.	—
Ha	27° 17.6'	Ho	27° 16'

MOON ONLY SEXTANT (ALMANAC)

Corr.		Hs
Add'l Corr.		IC
HP	Corr.	Dip
Total		Ha
		Total
		Ho

ASS. POSITION LHA (ALMANAC)

GHA Hours	
Min. & Sec.	
Stars Only SHA	
Moon & Planets Only V:	Corr.
360°	
Total GHA	
Ass. Long.	
LHA	281°

INTERCEPT AZIMUTH – FIRST ENTRY

Lat.	LHA	A	A°	A'
		B ±	Z_1 ±	
		Dec. ±		
	Sum: B+Dec.	F	F°	F'

PLOTTING

Ass. Long.		
Zn 281°	Ass. Lat.	
Intercept	T	A 15'

INTERCEPT AZIMUTH – SECOND ENTRY

A°	F°	H	P°	Z_2 ±
F'	P°	Corr.₁ ±	$Z_2°$	
	Sum		Z_1	
A'	$Z_2°$	Corr.₂ ±	Z_2	
	True Hc	27° 31'	Z	
	True Ho	27° 16'	Zn	
	Intercept	15'	T	A ✓

DECLINATION (ALMANAC)

Dec. Hours	
"d" ±	Corr. ±
True Dec.	

FIGURE 10.3 ☆

Go to Figure 10.5 to the page headed: LAT. 22° N from *Vol. I, Pub. No. 249, Sight Reduction Tables for Air Navigation.*

Inspecting this table that represents half of the information for Lat. 22° N, you observe that for all the possible LHAs of Aries, to the nearest degree, there are seven selected stars, complete with their tabulated altitudes (True Hc) and their true 360° azimuths (Zn).

What you now have is not only a star finder but calculations of altitude and azimuth that you heretofore calculated manually.

The three stars most suitable for a three-star fix are marked with a ♦ and the names of the stars of first magnitude are capitalized.

Of the 41 stars selected for publication, 19 stars are of the first magnitude.

My tables were for EPOCH 1985.0. If you decide to purchase these tables you will probably acquire a later epoch, the tabulated altitudes and azimuths will be slightly different.

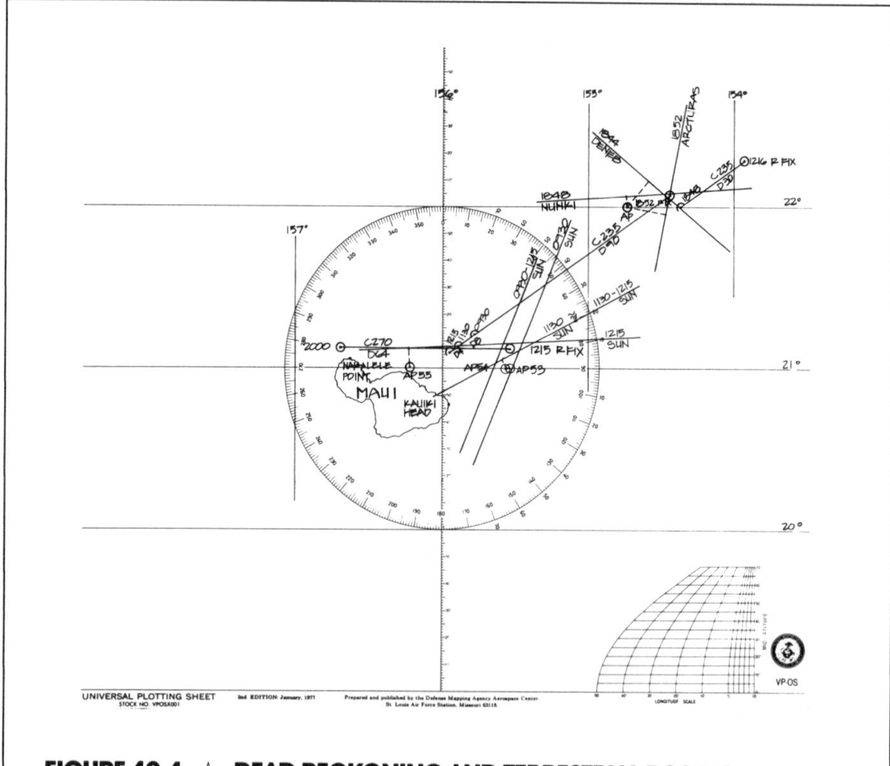

FIGURE 10.4 ☆ DEAD RECKONING AND TERRESTRIAL POSITIONS FROM 9/23/86 TO 9/24/86

Go down the column for the LHA of Aries to 279°.

Go across to the first star marked with a ♦ and record on the worksheet No. 50: Body: DENEB and True Hc: 55°31′.

In the Plotting block space record: Zn: 40°.

Go down the column for the LHA of Aries to 280°.

Go across to the second star with a ♦ and record on worksheet No. 51: Body: NUNKI and True Hc: 41°34′.

In the Plotting block space record: Zn: 176°.

Follow the same procedure for LHA 281° and record the tabulated information for ARCTURUS on worksheet No. 52.

We now have all the information required to take observations on Deneb, Nunki, and Arcturus.

LAT 22°N

LHA ♈	Hc Zn	Hc Zn	Hc Zn	Hc Zn	Hc Zn	Hc Zn	Hc Zn
	•Alkaid	Alphecca	•SPICA	Gienah	REGULUS	•POLLUX	Dubhe
180	55 25 031	41 22 073	50 59 146	50 22 174	61 24 254	32 25 290	48 58 350
181	55 54 030	42 16 073	51 29 147	50 27 176	60 30 255	31 33 290	48 48 349
182	56 22 029	43 09 073	51 59 149	50 30 177	59 37 256	30 41 291	48 37 348
183	56 49 029	44 02 073	52 27 150	50 32 179	58 43 257	29 49 291	48 26 348
184	57 15 028	44 56 074	52 54 151	50 32 180	57 48 257	28 57 291	48 14 347
185	57 40 027	45 49 074	53 20 153	50 31 182	56 54 258	28 05 291	48 01 347
186	58 05 026	46 42 074	53 45 154	50 29 183	56 00 259	27 13 291	47 48 346
187	58 29 025	47 36 074	54 08 156	50 25 185	55 05 259	26 21 291	47 34 345
188	58 52 024	48 29 074	54 30 157	50 19 186	54 10 260	25 30 292	47 19 345
189	59 14 023	49 23 074	54 51 159	50 13 188	53 15 260	24 38 292	47 04 344
190	59 35 022	50 16 074	55 10 161	50 04 189	52 20 261	23 46 292	46 49 344
191	59 55 021	51 10 074	55 28 162	49 55 191	51 25 262	22 55 292	46 33 343
192	60 14 019	52 03 074	55 44 164	49 43 192	50 30 262	22 03 293	46 16 342
193	60 32 018	52 57 074	55 59 166	49 31 194	49 35 263	21 12 293	45 59 342
194	60 49 017	53 50 074	56 11 167	49 17 195	48 40 263	20 21 293	45 42 341
	•Kochab	VEGA	Rasalhague	•ANTARES	SPICA	•REGULUS	Dubhe
195	35 43 009	17 59 055	24 21 086	20 05 131	56 23 169	47 45 264	45 24 341
196	35 52 009	18 45 055	25 17 086	20 47 132	56 32 171	46 49 264	45 05 340
197	36 00 008	19 30 055	26 12 086	21 28 132	56 40 173	45 54 265	44 46 340
198	36 08 008	20 16 055	27 08 087	22 09 133	56 47 174	44 59 265	44 27 339
199	36 16 008	21 01 055	28 03 087	22 49 134	56 51 176	44 03 266	44 07 339
200	36 23 007	21 47 056	28 59 087	23 29 134	56 54 178	43 08 266	43 46 338
201	36 30 007	22 33 056	29 54 088	24 09 135	56 55 180	42 12 266	43 26 338
202	36 37 007	23 19 056	30 50 088	24 48 136	56 54 182	41 17 267	43 05 338
203	36 43 007	24 05 056	31 46 089	25 27 136	56 52 183	40 21 267	42 43 337
204	36 50 006	24 51 056	32 41 089	26 05 137	56 48 185	39 26 268	42 21 337
205	36 55 006	25 38 056	33 37 089	26 42 138	56 42 187	38 30 268	41 59 336
206	37 01 006	26 24 056	34 32 090	27 20 138	56 34 189	37 34 268	41 37 336
207	37 06 005	27 10 057	35 28 090	27 56 139	56 25 191	36 39 269	41 14 336
208	37 11 005	27 57 057	36 24 090	28 32 140	56 14 192	35 43 269	40 51 335
209	37 16 005	28 43 057	37 19 091	29 08 141	56 01 194	34 48 270	40 27 335
	•Kochab	VEGA	Rasalhague	•ANTARES	SPICA	•REGULUS	Dubhe
210	37 20 004	29 30 057	38 15 091	29 43 142	55 47 196	33 52 270	40 03 334
211	37 24 004	30 16 057	39 11 092	30 17 142	55 31 197	32 56 270	39 39 334
212	37 28 004	31 03 057	40 06 092	30 51 143	55 14 199	32 01 271	39 15 334
213	37 31 003	31 50 057	41 02 092	31 24 144	54 55 201	31 05 271	38 50 334
214	37 34 003	32 36 057	41 57 093	31 56 145	54 35 202	30 09 271	38 25 333
215	37 37 003	33 23 057	42 53 093	32 28 146	54 13 204	29 14 272	38 00 333
216	37 40 002	34 10 057	43 48 094	32 59 147	53 50 205	28 18 272	37 35 333
217	37 42 002	34 57 057	44 44 094	33 29 147	53 25 207	27 23 273	37 09 332
218	37 43 002	35 43 057	45 39 094	33 59 148	53 00 208	26 27 273	36 43 332
219	37 45 001	36 30 057	46 35 095	34 28 149	52 33 210	25 31 273	36 17 332
220	37 46 001	37 17 057	47 30 095	34 56 150	52 05 211	24 36 274	35 51 332
221	37 47 001	38 04 057	48 26 096	35 23 151	51 35 213	23 40 274	35 25 332
222	37 47 000	38 50 057	49 21 096	35 50 152	51 05 214	22 45 274	34 58 331
223	37 47 000	39 37 057	50 16 097	36 15 153	50 33 215	21 49 275	34 31 331
224	37 47 000	40 24 057	51 11 097	36 40 154	50 01 216	20 54 275	34 04 331
	VEGA	•ALTAIR	Nunki	ANTARES	•SPICA	Denebola	•Alkaid
225	41 11 057	19 26 088	15 30 127	37 04 155	49 27 218	44 06 269	59 02 337
226	41 57 057	20 22 088	16 14 128	37 27 156	48 53 219	43 10 270	58 40 336
227	42 44 057	21 18 089	16 57 129	37 49 157	48 17 220	42 14 270	58 16 335
228	43 30 057	22 13 089	17 41 129	38 11 158	47 41 221	41 19 270	57 52 334
229	44 17 057	23 09 090	18 24 130	38 31 159	47 04 222	40 23 271	57 27 333
230	45 03 057	24 04 090	19 06 130	38 50 160	46 26 223	39 28 271	57 01 332
231	45 50 056	25 00 090	19 49 131	39 09 161	45 47 225	38 32 272	56 34 331
232	46 36 056	25 56 091	20 30 131	39 26 162	45 08 226	37 36 272	56 07 330
233	47 22 056	26 51 091	21 12 132	39 42 163	44 28 227	36 41 272	55 39 329
234	48 08 056	27 47 091	21 53 133	39 58 165	43 47 228	35 45 273	55 10 329
235	48 54 056	28 43 092	22 34 133	40 12 166	43 06 229	34 50 273	54 41 328
236	49 40 055	29 38 092	23 14 134	40 25 167	42 24 229	33 54 273	54 11 327
237	50 26 055	30 34 093	23 54 135	40 37 168	41 41 230	32 59 274	53 40 326
238	51 12 055	31 29 093	24 33 135	40 48 169	40 58 231	32 03 274	53 09 326
239	51 57 055	32 25 093	25 12 136	40 58 170	40 14 232	31 08 274	52 38 325
	•VEGA	ALTAIR	Nunki	•ANTARES	SPICA	ARCTURUS	•Alkaid
240	52 42 054	33 20 094	25 51 137	41 07 172	39 30 233	65 18 268	52 06 324
241	53 27 054	34 16 094	26 29 137	41 14 173	38 45 234	64 23 269	51 33 324
242	54 12 054	35 11 095	27 06 138	41 21 174	38 00 235	63 27 269	51 00 323
243	54 57 053	36 07 095	27 43 139	41 26 175	37 15 235	62 31 270	50 27 323
244	55 41 053	37 02 096	28 19 140	41 30 176	36 29 236	61 36 270	49 53 322
245	56 25 052	37 57 096	28 55 140	41 33 177	35 42 237	60 40 270	49 19 322
246	57 09 052	38 53 096	29 30 141	41 35 179	34 55 238	59 45 271	48 44 321
247	57 53 051	39 48 097	30 05 142	41 36 180	34 08 238	58 49 271	48 09 321
248	58 36 051	40 43 097	30 39 143	41 36 181	33 21 239	57 53 271	47 34 320
249	59 19 050	41 38 098	31 13 143	41 34 182	32 33 240	56 58 272	46 58 320
250	60 01 049	42 33 098	31 45 144	41 31 183	31 45 240	56 02 272	46 23 320
251	60 43 049	43 28 099	32 17 145	41 27 185	30 56 241	55 07 272	45 46 319
252	61 25 048	44 23 099	32 49 146	41 22 186	30 07 242	54 11 273	45 10 319
253	62 06 047	45 18 100	33 20 147	41 16 187	29 18 242	53 15 273	44 33 319
254	62 46 046	46 13 101	33 50 148	41 09 188	28 29 243	52 20 273	43 57 318
	•DENEB	ALTAIR	•Nunki	ANTARES	SPICA	•ARCTURUS	Alkaid
255	39 40 049	47 08 101	34 19 149	41 00 189	27 39 244	51 24 274	43 20 318
256	40 22 049	48 02 102	34 47 150	40 51 191	26 49 244	50 29 274	42 42 318
257	41 04 048	48 57 102	35 15 151	40 40 192	25 59 245	49 33 274	42 05 318
258	41 45 048	49 51 103	35 42 152	40 28 193	25 09 245	48 38 275	41 27 317
259	42 27 048	50 45 103	36 08 153	40 15 194	24 18 246	47 42 275	40 50 317
260	43 08 048	51 39 104	36 33 153	40 01 195	23 27 246	46 47 275	40 12 317
261	43 49 048	52 33 105	36 58 154	39 46 196	22 36 247	45 52 275	39 34 317
262	44 30 047	53 27 105	37 21 155	39 30 197	21 45 247	44 56 276	38 55 317
263	45 11 047	54 20 106	37 44 157	39 13 198	20 53 248	44 01 276	38 17 316
264	45 52 047	55 13 107	38 06 158	38 55 200	20 02 248	43 06 276	37 39 316
265	46 32 047	56 07 108	38 26 159	38 36 201	19 10 249	42 10 277	37 00 316
266	47 13 046	56 59 109	38 46 160	38 16 202	18 18 249	41 15 277	36 22 316
267	47 53 046	57 52 109	39 05 161	37 55 203	17 26 250	40 20 277	35 43 316
268	48 33 046	58 44 110	39 23 162	37 33 204	16 33 250	39 25 277	35 04 316
269	49 12 045	59 36 111	39 40 163	37 10 205	15 41 251	38 30 278	34 25 316

LAT 22°N

LHA ♈	Hc Zn	Hc Zn	Hc Zn	Hc Zn	Hc Zn	Hc Zn	Hc Zn
	•DENEB	Enif	•Nunki	ANTARES	•ARCTURUS	Alkaid	Kochab
270	49 52 045	35 12 093	39 55 164	36 46 206	37 34 278	33 46 316	32 06 346
271	50 31 044	36 08 094	40 10 165	36 22 207	36 39 278	33 07 316	31 53 346
272	51 10 044	37 03 094	40 24 166	35 56 208	35 44 278	32 28 316	31 40 346
273	51 48 044	37 59 095	40 36 167	35 30 209	34 49 279	31 49 315	31 26 346
274	52 26 043	38 54 095	40 48 169	35 03 210	33 54 279	31 10 315	31 12 346
275	53 04 043	39 50 096	40 58 170	34 35 211	32 59 279	30 31 315	30 58 345
276	53 42 042	40 45 096	41 08 171	34 06 211	32 05 280	29 52 315	30 44 345
277	54 19 041	41 40 097	41 16 172	33 37 212	31 10 280	29 13 315	30 30 345
278	54 55 041	42 36 097	41 23 173	33 06 213	30 15 280	28 34 315	30 16 345
279	55 31 040	43 31 097	41 29 175	32 36 214	29 20 280	27 55 315	30 01 345
280	56 07 039	44 26 098	41 34 176	32 04 215	28 25 281	27 16 315	29 47 345
281	56 42 039	45 21 098	41 37 177	31 32 216	27 31 281	26 37 316	29 32 345
282	57 17 038	46 16 099	41 40 178	30 59 217	26 36 281	25 58 316	29 17 344
283	57 51 037	47 11 100	41 41 179	30 25 218	25 42 282	25 19 316	29 02 344
284	58 24 036	48 06 100	41 41 180	29 51 218	24 47 282	24 40 316	28 47 344
	DENEB	•Alpheratz	FOMALHAUT	•Nunki	ANTARES	•Alphecca	Kochab
285	58 57 036	21 26 066	13 06 130	41 40 182	29 16 219	43 10 287	28 32 344
286	59 29 035	22 17 066	13 49 131	41 38 183	28 41 220	42 17 287	28 17 344
287	60 00 034	23 08 067	14 31 131	41 34 184	28 05 221	41 24 287	28 01 344
288	60 30 033	23 59 067	15 13 132	41 30 185	27 29 221	40 31 287	27 46 344
289	61 00 032	24 50 067	15 54 132	41 24 186	26 52 222	39 37 287	27 30 344
290	61 29 031	25 42 067	16 35 133	41 17 188	26 14 223	38 44 287	27 14 344
291	61 57 030	26 33 067	17 16 133	41 09 189	25 36 224	37 51 288	26 59 344
292	62 24 028	27 24 068	17 56 134	41 00 190	24 57 224	36 58 288	26 43 343
293	62 50 027	28 16 068	18 36 134	40 50 191	24 18 225	36 05 288	26 27 343
294	63 15 026	29 07 068	19 16 135	40 38 192	23 39 226	35 12 288	26 11 343
295	63 38 025	29 59 068	19 55 136	40 26 193	22 59 226	34 19 288	25 55 343
296	64 01 023	30 51 068	20 33 136	40 13 195	22 19 227	33 27 288	25 39 343
297	64 22 022	31 43 069	21 12 137	39 58 196	21 38 227	32 34 289	25 23 343
298	64 42 020	32 34 069	21 49 138	39 42 197	20 57 228	31 41 289	25 07 343
299	65 01 019	33 26 069	22 27 138	39 26 198	20 15 229	30 48 289	24 51 343
	•DENEB	Schedar	Alpheratz	•FOMALHAUT	Nunki	•Rasalhague	VEGA
300	65 19 017	29 13 036	34 18 069	23 04 139	39 08 199	54 03 261	65 30 318
301	65 34 016	29 46 036	35 10 069	23 40 140	38 49 200	53 08 262	64 52 317
302	65 49 014	30 19 036	36 02 069	24 16 140	38 30 201	52 13 262	64 14 316
303	66 02 013	30 52 036	36 54 069	24 51 141	38 09 202	51 18 263	63 35 315
304	66 13 011	31 25 036	37 46 070	25 26 142	37 48 203	50 23 263	62 55 314
305	66 23 009	31 58 036	38 39 070	26 00 142	37 25 204	49 27 264	62 14 313
306	66 31 007	32 31 036	39 31 070	26 34 143	37 02 205	48 32 264	61 33 312
307	66 37 006	33 04 036	40 23 070	27 07 144	36 37 206	47 37 265	60 52 311
308	66 42 004	33 36 036	41 15 070	27 40 145	36 12 207	46 41 265	60 10 311
309	66 45 002	34 09 036	42 08 070	28 12 145	35 46 208	45 46 265	59 28 310
310	66 46 000	34 41 036	43 00 070	28 43 146	35 20 209	44 50 266	58 45 309
311	66 46 359	35 13 035	43 52 070	29 14 147	34 52 210	43 55 266	58 02 309
312	66 44 357	35 46 035	44 45 070	29 43 148	34 24 211	42 59 267	57 18 308
313	66 40 355	36 18 035	45 37 070	30 13 149	33 55 212	42 04 267	56 35 308
314	66 34 353	36 49 035	46 29 070	30 41 149	33 25 213	41 08 268	55 51 307
	•Alpheratz	Diphda	•FOMALHAUT	ALTAIR	Rasalhague	•VEGA	DENEB
315	47 22 071	22 22 122	31 09 150	68 39 235	40 13 268	55 06 307	66 27 352
316	48 14 071	23 09 122	31 37 151	67 53 236	39 17 268	54 22 307	66 18 350
317	49 07 071	23 55 123	32 03 152	67 06 238	38 21 269	53 37 306	66 07 348
318	49 59 071	24 42 124	32 29 153	66 19 239	37 26 269	52 52 306	65 55 347
319	50 52 071	25 28 124	32 54 154	65 30 241	36 30 270	52 07 305	65 41 345
320	51 44 071	26 14 125	33 18 155	64 42 242	35 35 270	51 21 305	65 26 343
321	52 37 071	26 59 126	33 41 156	63 52 243	34 39 270	50 36 305	65 09 342
322	53 29 071	27 44 126	34 03 157	63 02 245	33 43 271	49 50 305	64 51 340
323	54 22 070	28 29 127	34 25 158	62 12 246	32 48 271	49 04 304	64 32 339
324	55 14 070	29 13 128	34 46 159	61 21 247	31 52 271	48 18 304	64 11 337
325	56 06 070	29 57 128	35 06 160	60 29 248	30 56 272	47 32 304	63 49 336
326	56 59 070	30 41 129	35 25 161	59 38 249	30 01 272	46 46 304	63 26 335
327	57 51 070	31 24 130	35 43 162	58 46 250	29 05 273	46 00 304	63 01 333
328	58 43 070	32 06 130	36 00 163	57 53 251	28 10 273	45 13 303	62 36 332
329	59 36 070	32 49 131	36 16 164	57 01 251	27 14 273	44 27 303	62 09 331
	•Mirfak	Hamal	Diphda	•FOMALHAUT	ALTAIR	•VEGA	DENEB
330	22 26 044	33 37 076	33 30 132	36 31 165	56 08 252	43 40 303	61 42 330
331	23 05 044	34 31 076	34 11 133	36 46 166	55 15 253	42 54 303	61 13 329
332	23 43 044	35 25 076	34 52 134	36 59 167	54 21 254	42 07 303	60 44 328
333	24 22 044	36 19 076	35 32 134	37 11 168	53 28 254	41 21 303	60 14 327
334	25 00 044	37 13 077	36 11 135	37 22 169	52 34 255	40 34 303	59 43 326
335	25 39 044	38 08 077	36 50 136	37 33 170	51 40 256	39 47 303	59 11 325
336	26 17 044	39 02 077	37 29 137	37 42 171	50 46 257	39 00 303	58 39 324
337	26 56 044	39 56 077	38 06 138	37 50 172	49 52 257	38 14 303	58 06 323
338	27 35 044	40 50 077	38 43 139	37 57 173	48 58 258	37 27 303	57 32 322
339	28 13 044	41 45 078	39 19 140	38 03 174	48 03 258	36 40 303	56 58 322
340	28 52 044	42 39 078	39 55 141	38 08 175	47 09 259	35 53 303	56 23 321
341	29 31 044	43 33 078	40 30 142	38 12 176	46 14 259	35 07 303	55 48 320
342	30 10 044	44 28 078	41 04 143	38 15 178	45 20 260	34 20 303	55 12 319
343	30 48 044	45 22 078	41 37 144	38 17 179	44 25 261	33 33 303	54 35 319
344	31 27 044	46 17 079	42 10 145	38 18 180	43 30 261	32 46 303	53 59 318
	Schedar	•CAPELLA	ALDEBARAN	Diphda	•FOMALHAUT	ALTAIR	•DENEB
345	50 59 022	13 02 045	11 42 077	42 41 146	38 18 181	42 35 262	53 21 318
346	51 19 021	13 42 046	12 36 077	43 12 147	38 16 182	41 40 262	52 44 317
347	51 38 020	14 22 046	13 31 077	43 42 148	38 14 183	40 45 263	52 06 317
348	51 57 020	15 02 046	14 25 078	44 11 149	38 10 184	39 49 263	51 27 316
349	52 16 019	15 42 046	15 19 078	44 39 150	38 06 185	38 54 263	50 49 316
350	52 33 018	16 22 046	16 14 078	45 06 152	38 00 186	37 59 264	50 10 315
351	52 50 017	17 02 047	17 08 079	45 32 153	37 53 187	37 04 264	49 30 315
352	53 06 016	17 43 047	18 03 079	45 57 154	37 45 189	36 08 265	48 51 315
353	53 22 016	18 24 047	18 58 079	46 21 155	37 37 190	35 13 265	48 11 314
354	53 36 015	19 04 047	19 52 080	46 43 157	37 27 191	34 17 266	47 31 314
355	53 50 014	19 45 047	20 47 080	47 05 158	37 16 192	33 22 266	46 51 314
356	54 03 013	20 26 047	21 42 080	47 25 159	37 04 193	32 26 267	46 10 313
357	54 15 012	21 07 048	22 37 081	47 45 160	36 51 194	31 31 267	45 30 313
358	54 27 011	21 48 048	23 32 081	48 03 162	36 37 195	30 35 267	44 49 313
359	54 37 010	22 29 048	24 27 081	48 20 163	36 22 196	29 40 268	44 08 312

117

FIGURE 10.5 ☆ PAGE FROM *VOL. I, PUB. NO. 249* (TYPICAL)

At about 4h 40m GMT we began listening to stations WWV and WWVH, and checked the sextant for index error. The mark on the index arm corresponding to the micrometer drum pointed to 58′. We turned up the volume on the receiver so I could hear the time signal and announcements topside.

Prior to each sight, I preset the sextant to the computed altitude, Hc, and peered over the binnacle compass in the direction of the star. When I observed the star in the sextant's horizon glass near the horizon, I moved to a convenient position for taking the sight and rotated the micrometer drum to keep the star on the horizon. When I heard the time signal for each four minute interval, I recorded the sextant reading and prepared for the next sight.

Record the altitudes: Hs: 55°46′; 41°33′; and 27°19′ respectively.

Calculate and record the dip: 11 feet for the first sight and 12 feet for the remaining sights.

Record the altitude corrections from the second column on the inside front cover of the almanac.

There were no temperature and pressure corrections.

After calculating Ho for each sight, transfer the Ho readings to True Ho on the worksheets.

Calculate the intercepts, plot the resultant LOPs from the same assumed position and label them accordingly.

There is only one small task remaining. Go to Figure 10.6. Toward the end of *Vol. I, Pub. No. 249, Sight Reduction Tables for Air Navigation* is a table headed: TABLE 5—CORRECTION FOR PRECESSION AND NUTATION. The 1985.0 table has no corrections for 1984 and 1985. For 1986 a small adjustment is used to correct the position of the fix.

Go down the first column headed: L.H.A. of Aries to 270° and go across under the heading: North latitudes to 20° N (a close enough approximation for LHA & Lat.). The first figure, a bold one, represents a one nautical mile displacement, and the second figure is the true azimuth of that displacement.

Note, that on our plot of this position, the fix was displaced just outside of the little triangle formed by the intersection of the three lines of position. This displacement was one nautical mile in the true direction: 100°. The position obtained was circled and labelled: 1852 FIX and the new DR course line was plotted in the direction 235° true.

It was especially nice to obtain a position of this accuracy as

TABLE 5—CORRECTION FOR PRECESSION AND NUTATION

L.H.A. ♈	North latitudes								South latitudes							L.H.A. ♈
	N. 89°	N. 80°	N. 70°	N. 60°	N. 50°	N. 40°	N. 20°	0°	S. 20°	S. 40°	S. 50°	S. 60°	S. 70°	S. 80°	S. 89°	
								1986								
°	′ °	′ °	′ °	′ °	′ °	′ °	′ °	′ °	′ °	′ °	′ °	′ °	′ °	′ °	′ °	°
0	0 —	0 —	1 030	1 040	1 050	1 060	1 060	1 070	1 070	1 060	1 060	1 050	1 050	1 030	0 —	0
30	0 —	1 030	1 050	1 050	1 060	1 060	1 070	1 070	1 060	1 060	1 050	1 040	1 030	0 —	0 —	30
60	0 —	1 060	1 060	1 070	1 070	1 070	1 070	1 070	1 070	1 060	1 050	0 —	0 —	0 —	0 —	60
90	0 —	1 080	1 080	1 080	1 080	1 080	1 080	1 080	1 080	1 080	0 —	0 —	0 —	0 —	0 —	90
120	0 —	1 100	1 100	1 100	1 100	1 100	1 100	1 100	1 100	1 100	0 —	0 —	0 —	0 —	0 —	120
150	0 —	1 120	1 120	1 110	1 110	1 110	1 110	1 110	1 110	1 120	1 130	0 —	0 —	0 —	0 —	150
180	0 —	1 150	1 130	1 130	1 120	1 120	1 110	1 110	1 120	1 120	1 130	1 140	1 150	0 —	0 —	180
210	0 —	0 —	1 150	1 140	1 130	1 120	1 120	1 110	1 110	1 120	1 120	1 130	1 130	1 150	0 —	210
240	0 —	0 —	0 —	0 —	1 130	1 120	1 110	1 110	1 110	1 110	1 110	1 110	1 120	1 120	0 —	240
270	0 —	0 —	0 —	0 —	0 —	1 100	1 100	1 100	1 100	1 100	1 100	1 100	1 100	1 100	0 —	270
300	0 —	0 —	0 —	0 —	0 —	1 080	1 080	1 080	1 080	1 080	1 080	1 080	1 080	1 080	0 —	300
330	0 —	0 —	0 —	0 —	1 050	1 060	1 070	1 070	1 070	1 070	1 070	1 070	1 060	1 060	0 —	330
360	0 —	0 —	1 030	1 040	1 050	1 060	1 060	1 070	1 070	1 060	1 060	1 050	1 050	1 030	0 —	360
								1987								
0	1 350	1 010	1 030	1 050	1 050	2 060	2 070	2 070	2 070	2 060	2 060	1 050	1 040	1 030	1 010	0
30	1 020	1 040	1 050	1 060	2 060	2 060	2 070	2 070	2 070	1 060	1 050	1 040	1 030	1 000	1 340	30
60	1 050	1 060	1 070	2 070	2 070	2 070	2 080	2 070	2 070	1 060	1 050	1 040	1 010	1 330	1 310	60
90	1 080	1 080	1 080	2 090	2 090	2 090	2 090	2 090	2 090	1 080	1 080	0 —	0 —	1 290	1 280	90
120	1 110	1 110	1 100	2 100	2 100	2 100	2 100	2 100	2 100	1 110	1 120	0 —	0 —	1 240	1 250	120
150	1 140	1 130	1 120	2 110	2 110	2 110	2 110	2 110	2 110	1 120	1 130	1 140	1 170	1 200	1 220	150
180	1 170	1 150	1 140	1 130	2 120	2 120	2 110	2 110	2 120	2 120	1 130	1 140	1 150	1 170	1 190	180
210	1 200	1 180	1 150	1 140	1 130	1 120	2 120	2 110	2 110	2 120	2 120	1 120	1 130	1 140	1 160	210
240	1 230	1 210	1 180	1 140	1 130	1 120	2 110	2 110	2 110	2 110	2 110	2 110	1 110	1 120	1 130	240
270	1 260	1 250	0 —	0 —	1 100	1 100	2 100	2 090	2 090	2 090	2 090	2 100	1 100	1 100	1 100	270
300	1 290	1 300	0 —	0 —	1 070	1 070	2 080	2 080	2 080	2 080	2 080	2 080	1 080	1 080	1 070	300
330	1 320	1 340	1 010	1 040	1 050	1 060	2 070	2 070	2 070	2 070	2 070	2 070	1 060	1 050	1 040	330
360	1 350	1 010	1 030	1 050	1 050	2 060	2 070	2 070	2 070	2 060	2 060	1 050	1 040	1 030	1 010	360
								1988								
0	1 000	1 020	1 030	2 050	2 050	2 060	3 070	3 070	3 070	3 060	2 060	2 050	2 040	1 030	1 010	0
30	1 030	1 040	2 050	2 060	2 060	3 070	3 070	3 070	3 070	2 060	2 050	1 040	1 020	1 000	1 340	30
60	1 060	2 060	2 070	2 070	3 070	3 080	3 080	3 080	2 070	2 060	1 060	1 040	1 000	1 330	1 310	60
90	1 080	2 090	2 090	2 090	3 090	3 090	3 090	3 090	2 090	1 080	1 080	0 —	0 —	1 280	1 280	90
120	1 110	2 110	2 100	2 100	3 100	3 100	3 100	3 100	2 100	1 110	1 120	1 140	0 —	1 230	1 250	120
150	1 140	2 130	2 120	2 120	3 110	3 110	3 110	3 110	3 110	2 120	2 130	1 140	1 160	1 190	1 220	150
180	1 170	1 150	2 140	2 130	2 120	3 120	3 110	3 110	3 120	2 120	2 130	2 130	1 150	1 160	1 180	180
210	1 200	1 180	1 160	1 140	2 130	2 120	3 110	3 110	3 110	3 120	2 120	2 120	2 130	1 140	1 160	210
240	1 230	1 210	1 180	1 140	1 130	2 120	2 110	3 110	3 100	3 110	3 110	2 110	2 110	2 120	1 130	240
270	1 260	1 260	0 —	0 —	1 100	1 100	2 090	3 090	3 090	3 090	3 090	2 090	2 090	2 100	1 100	270
300	1 290	1 310	0 —	1 040	1 060	1 070	2 080	3 080	3 080	3 080	3 080	2 080	2 080	2 070	1 070	300
330	1 330	1 350	1 020	1 040	2 050	2 060	3 070	3 070	3 070	3 070	3 070	2 060	2 060	2 050	1 040	330
360	1 000	1 020	1 030	2 050	2 050	2 060	3 070	3 070	3 070	3 060	2 060	2 050	2 040	1 030	1 010	360
								1989								
0	2 000	2 020	2 040	2 050	3 060	3 060	4 070	4 070	4 070	3 060	3 060	2 050	2 040	2 030	2 010	0
30	2 030	2 040	2 050	3 060	3 060	3 070	4 070	4 070	3 070	3 060	2 050	2 040	2 020	1 000	2 340	30
60	2 060	2 070	3 070	3 070	3 080	4 080	4 080	4 080	3 070	2 070	2 060	1 040	1 000	1 320	2 310	60
90	2 090	2 090	3 090	3 090	4 090	4 090	4 090	4 090	3 090	2 090	1 080	0 —	0 —	1 280	2 280	90
120	2 120	2 110	3 110	3 100	3 100	4 100	4 100	4 100	3 100	2 110	1 120	1 140	1 190	1 230	2 240	120
150	2 140	2 130	2 120	3 120	3 110	4 110	4 110	4 110	3 110	3 120	2 130	2 140	1 160	1 190	2 210	150
180	2 170	2 160	2 140	2 130	3 120	3 120	4 120	4 110	4 120	3 120	3 130	2 130	2 140	2 160	2 180	180
210	2 200	1 180	2 160	2 140	2 130	3 120	3 110	4 110	4 110	3 110	3 120	3 120	2 130	2 140	2 150	210
240	2 230	1 220	1 180	1 140	2 120	2 120	3 110	4 100	4 100	4 100	3 110	3 110	3 110	2 120	2 120	240
270	2 270	1 260	0 —	0 —	1 100	2 090	3 090	4 090	4 090	4 090	4 090	3 090	3 090	2 090	2 090	270
300	2 300	1 310	1 350	1 040	1 060	2 070	3 080	4 080	4 080	4 080	3 080	3 080	3 080	2 070	2 070	300
330	2 330	1 350	1 020	2 040	2 050	3 060	3 070	4 070	4 070	4 070	3 070	3 060	2 060	2 050	2 040	330
360	2 000	2 020	2 040	2 050	3 060	3 060	4 070	4 070	4 070	3 060	3 060	2 050	2 040	2 030	2 010	360

Example. In 1988 a position line is obtained in latitude S.52° when L.H.A. ♈ is 327°. Entering the table with the year 1988, latitude S.50°, and L.H.A. ♈ 330° gives 3 070° which indicates that the position line is to be transferred 3 miles in true bearing 070°.

323

FIGURE 10.6 ☆ *VOL. I, PUB. NO. 249*, CORRECTION FOR PRECESSION AND NUTATION (TYPICAL)

we approached Maui. Because of our progress, we had every reason to believe that we would be making a landfall the next day. The next morning we attempted to obtain another star fix prior to sunrise, but were not successful because of excessive cloud cover and we had to wait for the sun to appear.

ARRIVAL - LAHAINA, MAUI, HAWAII

ELEVEN

THE LANDFALL

Many times in the past I have asked myself these questions: Why do I go to sea? Why have I not remained in a secure occupation? Why have I not purchased a home ashore? Why haven't I directed my efforts toward the accumulation of a personal fortune? The list is endless. I have quit asking myself these questions because they have all been answered in the process of living.

My wife and I have become aware that the most important pursuit for our sanity and personal well-being is to follow our dreams. One of our dreams is to see as much of the world as possible. We both thoroughly enjoy travelling and believe that sailing a yacht around the world is one of the best ways to achieve our goal. One of the most exciting parts of the dream is about to unfold.

Today is a very special day. We have been at sea for 24 days, and are anticipating one of the greatest moments in the life of a blue water sailor . . . a landfall on a tropical island.

Once a landfall has been made, the responsibilities of navigation and seamanship intensify. There is always a feeling of relief and accomplishment, while a very important phase in navigation has just begun. You will usually be in waters you are not familiar with, and your knowledge of coastal piloting will be of special importance.

There are countless examples of extremely well-executed voyages that have ended in tragedy. Recently, a beautiful cutter terminated her passage from California to Hawaii on the reef at Diamond Head. Although no one was injured, the yacht was completely destroyed by waves grinding the hull into the reef. The accident was particularly difficult to understand, as the shore near Diamond Head has many navigational aids for mariners. The navigable limit of the point is buoyed, there is a lighthouse well known as the finish line for the Transpacific Yacht Race, and a well-lighted shoreline complete with many prominent land features. There were no problems with visibility and there was a full moon.

There is an extremely important rule that should be burned into the consciousness of every small craft navigator. It is, simply: Do not, under any circumstances, close with land after dark.

Earlier I mentioned that we sailed into Lahaina on the island of Maui. Initially we had planned to sail into Hilo on the big island of Hawaii. A few days ago, however, we determined by our daily progress that we would probably make a landfall off the big island in the middle of the night. If darkness were the only factor we would have hove-to and ensured a daytime landfall. But the weather was turning bad.

We changed our course from Hawaii to Maui because we were receiving weather reports forecasting heavy cloud cover, diminishing visibility, and some indication of tropical cyclonic activity on the Hamakua Coast of the Big Island. All things considered, we determined that it would be advantageous to plan a daytime landfall on Maui while maintaining our way.

When making a landfall, many different sightings over a shorter period of time are recommended. . . if possible, star fixes and at least three sun sights to obtain a position. The sun sights may be the best the weather will allow. If the sun sights were the only sights obtainable, I would not be disappointed. My philosophy has become: "Any position is better than no position, and I will be grateful."

Three sun sights are exactly what we were able to obtain. It was not possible at twilight that morning to determine our position. The cloud cover completely blanketed the area and there were no stars visible to sight. Toward mid-morning the sun started to appear intermittently through the clouds. Three sights were taken over a period of almost three hours as shown in Figures 11.1, 11.2 and 11.3.

We began plotting our DR position on both the plotting sheet and the Hawaiian Islands chart that morning, and transferred our newly obtained position from the plotting sheet to the chart.

You may reduce the sights if you wish. The only difference in technique, for taking these sights, was that a stopwatch was not started until the command "mark" was given. The stopwatch was

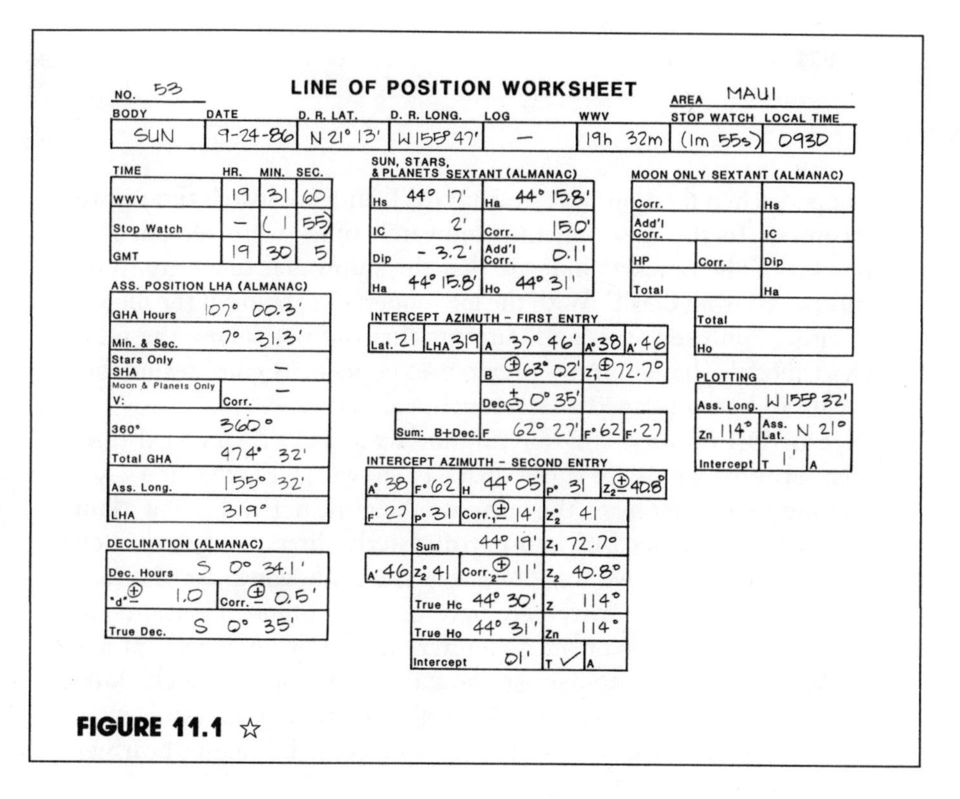

NO. 53 LINE OF POSITION WORKSHEET AREA MAUI

BODY	DATE	D. R. LAT.	D. R. LONG.	LOG	WWV	STOP WATCH	LOCAL TIME
SUN	9-24-86	N 21° 13'	W 155° 47'	—	19h 32m	(1m 55s)	0930

TIME

	HR.	MIN.	SEC.
WWV	19	31	60
Stop Watch	—	(1	55)
GMT	19	30	5

SUN, STARS, & PLANETS SEXTANT (ALMANAC)

Hs	44° 17'	Ha	44° 15.8'
IC	2'	Corr.	15.0'
Dip	- 3.2'	Add'l Corr.	0.1'
Ha	44° 15.8'	Ho	44° 31'

MOON ONLY SEXTANT (ALMANAC)

Corr.		Hs
Add'l Corr.		IC
HP	Corr.	Dip
Total		Ha
Total		
Ho		

ASS. POSITION LHA (ALMANAC)

GHA Hours	107° 00.3'
Min. & Sec.	7° 31.3'
Stars Only SHA	—
Moon & Planets Only V: / Corr.	—
360°	360°
Total GHA	474° 32'
Ass. Long.	155° 32'
LHA	319°

DECLINATION (ALMANAC)

Dec. Hours	S 0° 34.1'		
"d" ±	1.0	Corr. ±	0.5'
True Dec.	S 0° 35'		

INTERCEPT AZIMUTH – FIRST ENTRY

Lat. 21	LHA 319	A 37° 46'	A° 38	A' 46	
		B ⊕63° 02'	Z_1 ⊕72.7°		
		Dec ± 0° 35'			
	Sum: B+Dec.	F 62° 27'	F° 62	F' 27	

INTERCEPT AZIMUTH – SECOND ENTRY

A° 38	F° 62	H 44° 05'	P° 31	Z_2 ⊕40.8°
F' 27	P° 31	Corr.₁ ± 14'	Z_2° 41	
		Sum 44° 19'	Z_1 72.7°	
A' 46	Z_2° 41	Corr.₂ ± 11'	Z_2 40.8°	
		True Hc 44° 30'	Z 114°	
		True Ho 44° 31'	Zn 114°	
		Intercept 01'	T ✓	A

PLOTTING

Ass. Long.	W 155° 32'		
Zn 114°	Ass. Lat. N 21°		
Intercept	T 1'	A	

FIGURE 11.1 ☆

NO. 54 **LINE OF POSITION WORKSHEET** AREA MAUI

BODY	DATE	D. R. LAT.	D. R. LONG.	LOG	WWV	STOP WATCH	LOCAL TIME
SUN	9-24-86	N 21° 08'	W 155° 54'	—	21h 32m	(1m 52s)	1130

TIME	HR.	MIN.	SEC.
WWV	21	31	60
Stop Watch	–	(1	52)
GMT	21	30	8

SUN, STARS, & PLANETS SEXTANT (ALMANAC)

Hs	65° 34'	Ha	65° 32.6'
IC	2'	Corr.	15.5'
Dip	– 3.4'	Add'l Corr.	—
Ha	65° 32.6'	Ho	65° 48'

MOON ONLY SEXTANT (ALMANAC)

Corr.		Hs	
Add'l Corr.		IC	
HP	Corr.	Dip	
Total		Ha	
		Total	
		Ho	

ASS. POSITION LHA (ALMANAC)

GHA Hours		137° 00.7'
Min. & Sec.		7° 32.0'
Stars Only SHA		—
Moon & Planets Only V:	Corr.	—
360°		360°
Total GHA		504° 33'
Ass. Long.		155° 33'
LHA		349°

DECLINATION (ALMANAC)

Dec. Hours			S 0° 36.0'
"d" ±	1.0	Corr. ±	0.5'
True Dec.			S 0° 37'

INTERCEPT AZIMUTH – FIRST ENTRY

Lat. 21	LHA 349	A 10° 16'	A° 10	A' 16
		B ± 68° 39'	Z_1 ± 86.0°	
		Dec ± 0° 37'		
	Sum: B+Dec.	F 68° 02'	F° 68	F' 02

PLOTTING

Ass. Long.	W 155° 33'		
Zn	153°	Ass. Lat.	N 21°
Intercept	T	A	4'

INTERCEPT AZIMUTH – SECOND ENTRY

A° 10	F° 68	H 65° 56'	P° 65	Z_2 ± 66.7
F' 2	P° 65	$Corr._1$ ± 02'	$Z_2°$ 67	
	Sum	65° 58'	Z_1 86.0°	
A' 16	$Z_2°$ 67	$Corr._2$ ± 06'	Z_2 66.7°	
	True Hc	65° 52'	Z 153°	
	True Ho	65° 48'	Zn 153°	
	Intercept	04'	T	A ✓

FIGURE 11.2 ☆

stopped when the tone was heard; GMT and stopwatch time were recorded. In this case, all the stopwatch times were recorded in brackets. The brackets indicate that the stopwatch times are to be subtracted from GMT. With the log inoperative, being at the mercy of large, confused seas, and currents that did not follow the pilot chart models, based on my previous experience, I consider us fortunate to have obtained a terrestrial position.

Our DR position and our running fix at 1215 hours local time were approximately 20 miles apart as shown in Figure 10.4. According to the *pilot chart*, the currents along the north shore of Maui would have displaced us in a northwesterly direction as indicated by our DR. Later, we were told that the trade winds for the past two weeks had been extremely light or nonexistent. The no-wind situation appeared to create a counter current or no current at all.

We changed our course, as shown, and at approximately 1300 the slopes of Haleakeala, the 10,000-foot mountain on Maui, were sighted. We verified our position and progress by taking bearings

NO. 55 **LINE OF POSITION WORKSHEET** AREA MAUI

BODY	DATE	D. R. LAT.	D. R. LONG.	LOG	WWV	STOP WATCH	LOCAL TIME
SUN	9-24-86	N 21° 06'	W 155° 57'	—	22h 15m	(2m 11s)	1215

TIME	HR.	MIN.	SEC.
WWV	22	14	60
Stop Watch	–	(2	11)
GMT	22	12	49

ASS. POSITION LHA (ALMANAC)

GHA Hours	152° 00.9'	
Min. & Sec.	3° 12.3'	
Stars Only SHA	—	
Moon & Planets Only V:	Corr.	—
360°	360°	
Total GHA	515° 13'	
Ass. Long.	156° 13'	
LHA	359°	

DECLINATION (ALMANAC)

Dec. Hours	S 0° 37.0'		
"d" ±	1.0	Corr. ±	0.2'
True Dec.	S 0° 37'		

SUN, STARS, & PLANETS SEXTANT (ALMANAC)

Hs	68° 01'	Ha	67° 59.6'
IC	2'	Corr.	15.6'
Dip	-3.4'	Add'l Corr.	—
Ha	67° 59.6'	Ho	68° 15'

INTERCEPT AZIMUTH – FIRST ENTRY

Lat. 21	LHA 359	A 0° 56'	A° 1	A' 56
		B ⊕ 69° 00'	Z₁ ⊕ 89.6°	
		Dec ± 0° 37'		
Sum: B+Dec.		F 68° 23'	F° 68	F' 23

INTERCEPT AZIMUTH – SECOND ENTRY

A° 1	F° 68	H 67° 59'	P° 87	Z_2 ⊕ 87.5
F' 23	P° 87	$Corr._1$ ⊕ 23'	$Z_2°$ 88°	
	Sum	68° 22'	Z_1 89.6°	
A' 56	$Z_2°$ 88	$Corr._2$ ⊕ 0'	Z_2 87.5°	
	True Hc	68° 22'	Z 177°	
	True Ho	68° 15'	Zn 177°	
	Intercept	07'	T	A ✓

MOON ONLY SEXTANT (ALMANAC)

Corr.		Hs
Add'l Corr.		IC
HP	Corr.	Dip
Total		Ha
		Total
		Ho

PLOTTING

Ass. Long.	W 156° 13'	
Zn 177°	Ass. Lat. N 21°	
Intercept	T	A 7'

FIGURE 11.3 ☆

from prominent features ashore with a hand-bearing compass. Other bearings were taken on a local radio station with the radio direction finder. The remainder of the day was spent sailing the North Shore of Maui.

We broke out the chart for the channel details between Lanai, Maui, and Molokai.

Shortly after sunset, the light off the extreme northern portion of the island was seen and left well to port. If this light had not been seen, we would have stood farther offshore, and considered several alternatives:

1. If, and only if, the shores of Maui and Molokai were clearly visible, we would make a turn to port and sail down the middle of the channel between them.
2. Sail well off the coast of Molokai and enter the channel between Molokai and Oahu at dawn.
3. Heave to, keeping the boat well to seaward until dawn.

4. Keep an open mind to any number of alternatives based on changing conditions.

After leaving the light on the northernmost part of Maui well astern, the light off the extreme western portion of Maui was sighted. We slowly changed our course to port and sailed down the middle of the 12-mile-wide channel between Maui and Molokai. Sailing the middle of the channel was not difficult because both coasts were marked with the lights from houses and automobiles on the peripheral roads of both islands.

This is the time when most mainland sailors, unaccustomed to reefs, get into trouble. We were sailing at night and our depth perception was less reliable in darkness. We consistently obtained bearings to verify our approximate position between these islands. At no time did we purposely favor either side of the channel.

We broke out the "Approaches to Lahaina" chart so that we would have the most detailed information possible on our destination. The voyage ended at 0100 local time on September 25, 1986, as the ZOE was anchored off Lahaina, in the roadstead famous from the days of the whaling ships. We were careful to anchor outboard of the other cruising yachts sharing the anchorage. If these yachts had not been there, we would have waited until dawn to anchor because we did not want to risk running onto the reef.

Why didn't we sail into a port on the windward side or North Shore of Maui? Because the trade winds create large waves. These waves build up as they travel over thousands of miles of open ocean. The two northern ports on Maui have significant problems because of them.

Hana, the first, is a very small anchorage with limited turning radii, tremendous surge, and an old broken-down dock that is used primarily by very small fishing craft. One mistake and you have lost your yacht. No thanks! The other port is right in the bight of Maui between two large mountains and is so windy that it is extremely uncomfortable. This is the barge harbor at Kahului. Even though Kahului has a large breakwater, the surge is terrible, and there are no facilities for yachts. Landing a dinghy is a sizeable task. Launching the dinghy and rowing against the wind when returning to your craft would probably be worse. The message is that leeward ports, sheltered from the tradewinds by mountains, are the most desirable for small craft.

Fortunately we both received plenty of rest during the last few days of our voyage. Getting enough rest was important because the last hours of the voyage were extremely wearing. After we were satisfied the anchor was holding, we hoisted the kerosene anchor light on the forestay, and I slept like a dead man for about six hours.

At 0700 we weighed the anchor and powered the ZOE into the harbor at Lahaina. We checked in with the harbor master and were able to rent space in the harbor for a week.

The first day ashore, after a long passage, is always wonderful. Lahaina is a beautiful little town, with great views of the islands of Kahoolawe, Lanai and Molokai. We spent the day walking around town, drinking fresh tropical juices, and eating everything in sight.

STEPPING THE MAST,
MORRO BAY, CA

TWELVE

THE SAILINGS I

To discuss the sailings and planning of our cruise to the Hawaiian Islands, it is necessary to go back to a time prior to our departure, a beautiful July day on the Central California Coast. The ZOE was almost ready for her passage to Hawaii. Over the last year I had repowered her with a new diesel engine, replaced the standing rigging, replaced chainplates, installed a tri-color running light on the masthead, had the liferaft repacked, and done many other chores.

The list of things to do had shrunk to manageable proportions, with provisioning and passage planning remaining. We planned to have enough food and water for ninety days though the passage would most likely be much shorter. It may be true that it is possible to catch fish and devise a water catchment system; however, I've been on passages where fishing was bad and there wasn't enough water to catch.

One of the most important aspects of planning your course is to insure that you will not be sailing too close to obstructions. Examples of obstructions would be reefs, pinnacles, and other landfalls not planned. Obstructions are another important reason to keep track of your DR position and plot your daily terrestrial position on the chart of the area sailed. Do not assume that you have passed an obstruction that you have not seen. As a safety factor, always allow considerably more sea room than you would require if set by the maximum current. I like to allow 50 miles beyond the set when passing an obstruction after dark.

Recall that we were twenty miles from our DR position when we determined our actual position off Maui. The distance of twenty miles had accumulated overnight, and was significant. I had used the information pertaining to the usual currents and my past experience in making a landfall off Maui. I hope I have made my point.

One of the nice things about sailing from Morro Bay, California, to the Hawaiian Islands, however is that there are no obstructions en route.

There is published material that I have found helpful in planning voyages of this scope. My favorite book is *NP 136, Ocean Passages for the World.* I have a copy of the third edition published in 1973. *Ocean Passages for the World* is published by The Hydrographer of the Navy (British) and can be obtained from chart stores that market Admiralty Charts or from marine mail order sources.

Ocean Passages contains *large*- and *small*-distance sailing routes all over the world, for different times of the year, and uses seasonal information on currents, weather, and wind for the determination of these routes.

I also have a copy of the *Sailing Directions for the North Pacific Ocean* published by the Defense Mapping Agency of the U.S. Government. The *Sailing Directions for the North Pacific Ocean* is a very detailed publication and has an incredible amount of information on the vast area it covers.

Over the years I have collected *pilot charts* for each month in the year for the North Pacific Ocean. The pilot charts show the frequency of gales, storm tracks, wind roses, currents, sea temperature, and other useful information.

For coastal information, I consult *Coast Pilot*, *Light List*, and *Radio Navigation Aids*, published by the Defense Mapping Agency of the U.S. Government. The *Coast Pilot*, *Light List*, and *Radio Navigation Aids* publications can be found at your local chart store.

The time most favored by yachtsmen for sailing to the Hawaiian Islands from California is May through July, July being the most favored month for the best overall sailing conditions. The hurricane season in the North Pacific Ocean is from late May to mid November. The worst months for hurricanes are August through October.

I would have preferred making our passage in July; however, Enid was graduating from school toward the end of August. So we decided to sail August 31, 1986, weather permitting, and not wait until next year. I figured, that if we encountered a tropical cyclone en route, that the storm would move in a northwesterly direction, losing strength along its route out of the tropics.

Therefore, I determined that we would plan to sail a *great circle course*. A great circle course from Morro Bay to Hilo would be the shortest and most northerly route to the Big Island of Hawaii. By sailing this course we would be above the tropics until our last few days at sea.

I estimated that the wind en route would probably be fairly light, and that this late in the season the passage would take longer than it would during the summer months. The trip in fact took several days longer than we anticipated, a total of twenty-four and a half days.

We experienced a dying tropical cyclone about two thirds of the way across. One evening, shortly after sunset, the wind shifted rapidly to the southeast. The change caused the wind to be well ahead of the beam. The wind blew approximately thirty knots and persisted for a brief six hours. I wonder what the wind velocity might have been if we had been sailing 200 miles further south. We tracked other major storms by listening, twice a day, to the hourly weather reports transmitted on stations WWV and WWVH.

There were several storms of sizable wind velocities; however, they all diminished as they turned to the north.

In May 1976, when we sailed to Hawaii from San Francisco, it was necessary for us to sail in a southwesterly direction until reaching the tradewinds. After we sailed into the northern extremes of the Northeast Tradewinds, we changed our course and sailed a great circle course for the remainder of the passage. Except for slight course changes to take advantage of better wind conditions, the great circle course was approximated by sailing a series of *rhumb line* courses to our destination.

Let's begin by discussing rhumb lines. A rhumb line is nothing more than a straight line on a Mercator Projection or *Mercator Chart*. Nautical charts, with the exception of special projections, are Mercator Projections. What this means is that a course called a *Mercator sailing* (rhumb line) can be drawn as a straight line from one point to the next on a nautical chart and the bearing of the rhumb line will be the same at every meridian the line crosses.

In our case, I drew a line from Morro Bay to Hilo across the North Pacific Ocean chart and measured the bearing with a protractor from one of the meridians. The bearing was 246° true. I had no intention of sailing this longer course; however, the rhumb line gave me an idea of what the mid-range courses would be on the great circle course I would calculate. I said the rhumb line is a longer course.

If the rhumb line was plotted on the surface of the earth it would curve south of the shortest distance between the departure point and the destination point. If a great circle course and a rhumb line were plotted describing a course in both hemispheres, the great circle course would curve north of the rhumb line north of the equator, intersect with the rhumb line at the equator and curve south of the rhumb line south of the equator.

Examples of great circles, on the earth's surface, are all the meridians and the equator. The latitudes north and south of the equator are not great circles and are called *small circles*. Great circle sailings are most useful for sailing in higher latitudes, east-west routes, and over longer distances. If you are sailing down a meridian or along the equator your course would describe both a great circle sailing and a Mercator sailing.

Sailing a great circle course from Morro Bay to Hilo would save us less than 20 nautical miles. Much more important, however, was the fact that the actual great circle course maximum distance north of the rhumb line was approximately 90 miles and was far enough north to avoid most tropical cylone activity.

To plot the great circle route we will be using the sight reduction tables contained in the *Nautical Almanac* as follows:

1. Latitude page becomes departure latitude.
2. Declination becomes destination latitude.
3. The LHA becomes the difference between departure longitude and destination longitude.
4. The calculated altitude becomes a factor for determining the distance from the point of departure to the destination.
5. The true azimuth becomes the initial course.

Morro Bay	Hilo	Difference in Long.
Lat. 35° N	Lat. 20° N	
Long. 121° W	Long. 155° W	DLo. 34°

Minutes of latitude and longitude are ignored because we are not concerned with course accuracy of less than 1°.

Open the almanac to the page headed: LATITUDE/A: 30°–35°.

Go down the LHA column to 34° and across to 35°. Obtain: A: 27°16′, B: + 49°49′, etc., continuing as with any sight reduction problem. Then use the following to calculate the distance to our destination and the initial course:

$D = (90° - {}^{*}Hc) \times 60$ = Distance to destination in nautical miles.

1. *Hc must be in degrees and fractions of degrees only or:
2. After subtracting whole number of degrees, multiply the degrees and add the minutes.

By the first method:

$Hc = 56°33'$ or dividing 33′ by 60 we obtain .55°. That is, $Hc = 56.55°$ and $D = (90° - 56.55°) \times 60 = 33.45° \times 60 = 2007'$ or 2007 nm (nautical miles).

By the second method:

$90° - Hc = 89°60' - 56°33' = 33°27'$

$D = 33° \times 60 + 27' = 1980' + 27' = 2007'$ or 2007 nm.

Since the course is the true azimuth:

$C = Zn = 360° - Z = 360° - 108° = 252°$ or C252 and the initial course has been determined.

I plotted the course from Morro Bay to the 125° meridian on the North Pacific chart and labelled the top: C252.

Forty-eight hours later I figured a leg of the course from 34°N, 125°W to the destination. The course became 250° and the distance to Hilo became 1801 nautical miles. See if you obtain the same answer.

To complete the great circle plot to Hilo you would repeat this procedure for each 5° of longitude until the 150° meridian.

At the 150° meridian you would draw the course to Hilo and measure the final course off the chart. Each of these calculated routes are nothing more than small segments of rhumb lines approximating the great circle course. It is not practical to calculate too many way-points. The course changes slightly at each way-point, thereby creating a lot of unneccesary work.

To see how the actual voyage progressed go to Figure 12.1 headed: GREAT CIRCLE SAILING USING THE NAUTICAL ALMANAC SIGHT REDUCTION TABLE.

I designed this form to update the course after our position was determined each day.

During the first few days we sighted the sun, Polaris, Altair, Antares, and Arcturus to calculate our daily position. Each time our position was determined we updated the great circle course and plotted a new course. The new course was obtained by using the almanac and the distance to the destination was calculated.

To steer by the new course, we had to correct for variation, deviation, current set, and other factors.

Recalculate this data from the plotting sheets that you have completed and check your results.

The variation was taken from the North Pacific chart, and the current from the pilot charts. The compass had 0° deviation within the range of courses we sailed on the great circle course.

It is important to realize that you would not attempt sailing back to the initially planned course, and that the daily position obtained by celestial navigation is treated as the new departure point for the ultimate destination. If there were obstructions along the planned course, the obstructions would be taken into consideration and avoided with all new course headings.

As previously mentioned, we changed our course from Hilo to Lahaina during the last few days of our passage. The distance remaining to our destination was less than 500 miles. We do not calculate a great circle course for the last 500 miles of a passage.

GREAT CIRCLE SAILING
USING THE NAUTICAL ALMANAC
SIGHT REDUCTION TABLE

DATE	DEPARTURE LAT.	DEPARTURE LONG.	DESTINATION LAT.	DESTINATION LONG.	LHA	DISTANCE TO DESTINATION	TRUE COURSE	MAG COURSE	DEV. COR COURSE	CURRENT LEEWAY ETC. (STEER THIS COURSE)
8/31/86	35°	121°	20°	155°	34°	2007	252°	237°	237°	243°
9/1	35°	123°	↓	↓	32°	1914	250°	235°	235°	245°
9/2	34°	125°	↓	↓	30°	1801	250°	235°	235°	243°
9/3	33°	127°	↓	↓	28°	1686	249°	234°	234°	242°
9/4	33°	129°	↓	↓	26°	1592	247°	232°	232°	240°
9/21	24°	151°	21°	157°	SAILED RHUMBLINE ↓	342	240°	228°	228°	230°
9/22	23°	152°	↓	↓	↓	272	240°	228°	228°	230°
9/23	22°	154°	↓	↓	↓	167	235°	223°	223°	223°
9/23	22°	154°	↓	↓	↓	142	235°	223°	223°	223°
9/24	21°	156°	↓	↓	↓	64	267°	255°	255°	260°

FIGURE 12.1 ☆

Instead, we drew a straight line on the chart from our last position to approximately the middle of Maui and sailed the rhumb line to our final destination. After determining our daily position, we changed the course as necessary toward our final destination, and not the initially planned rhumb line.

Another important consideration is the wind direction. We don't like to work hard, we try to keep dry, and we make every effort to conserve our energy. Whenever possible, we sail a ways upwind of our destination, so that we can complete the voyage reaching or running downwind in comfort.

THIRTEEN SOME WRINKLES

SELECTING AN ASSUMED POSITION

We have been working with the concept of an assumed position (AP) as the basis for determining our actual position via the practice of celestial navagation. The AP is selected to be as close as possible to the DR position. The AP is also selected to facilitate the use of tabulated solutions contained in the sight reduction tables.

One factor in determining the accuracy in the plotting of an LOP is that the intercept be as short as possible. The approximate limit established for the length of an intercept is 30 nautical miles. If a sight has been taken accurately, the AP is selected near to the actual position, the sight has been properly reduced, and the LOP has been carefully plotted, it is likely that the intercept will be within the 30 nautical mile limit.

There are situations that can interfere with finding a daily terrestrial position in a small craft. I have personally experienced some debilitating seasickness. On one trip, my "mal de mer" persisted for about five days. I was sailing alone from Honolulu, Hawaii to Santa Barbara, California, wasn't concerned with obstructions, and I knew I could determine my terrestrial position later. I short sailed and maintained a record of courses and distance travelled, so that I could construct a DR plot when I felt better. When the sickness passed I plotted the DR position and began to navigate.

I was pleasantly surprised when I discovered that my DR position was within thirty-five nautical miles of my terrestrial position. The first LOP I calculated was from a sun sight around mid-morning. The azimuth pointed in a southeasterly direction, the intercept was less than 5 nautical miles, and was away from the sun. The second LOP was calculated from a sun sight shortly before noon. The azimuth pointed a few degrees less than south, the intercept was greater than 30 nautical miles, and was away from the sun.

Because the intercept was greater than 30 miles I decided to select a different AP and rework the sight. The determination of a new AP did not present any difficulties and I was pleased because the new sight showed that I had made better progress on my course than expected.

The key to determining the new AP was the length of the intercept and which way it pointed. In this case, the intercept of the southerly azimuth pointed away. This meant that the new assumed position would be further north. I selected a new AP by adding one degree of latitude and proceeded to rework the sight. The azimuth of the reworked sight was about the same. However, the new intercept was less than 30 miles and in the toward direction.

I worked out a third sight that afternoon. Again, the LOP had an intercept within the 30 nautical mile limitation. After the first and second LOPs were advanced to the time of the last sight, I was able to determine a running fix.

As an exercise, you could change the APs of sights you have previously reduced. By changing an AP and reworking a sight you would soon realize that the new azimuth and intercept would lead you back to the initial AP. As experience is gained, you will be able to see when a computed altitude will yield an intercept of over 30 nautical miles, and change the AP prior to calculating an intercept.

We have discussed changing the latitude. If an azimuth is in an east-west direction you would change the LHA by adding or subtracting from the assumed longitude. Occasionally it is necessary to change both the latitude and longitude of the assumed position if an azimuth is northwest, northeast, southwest, or southeast.

AVERAGING SIGHTS

In the preceding chapters we have commented on the accuracy of taking sights. One method to insure an accurate sight is to take several sights, timed as close to each other as possible, and reduce the sights to determine their azimuths and intercepts. If all the sights are accurate, the azimuths and intercepts would be very close to each other. If a sight in the group is not taken accurately, the difference between it and the other sights will be significant. The following is an example of this technique:

Sight 1: Zn = 165°; Intercept: T = 6′
Sight 2: Zn = 165°; Intercept: T = 5′
Sight 3: Zn = 166°; Intercept: T = 7′
Sight 4: Zn = 166°; Intercept: A = 10′
Sight 5: An = 167°; Interdept: T = 6′

Sight 4 should be eliminated or reworked. If the result is calculated without error, the mistake is in observing the altitude, and the sight should be rejected.

The other four sights should be averaged, and the LOP plotted from the average azimuth and intercept. The method is as follows:

$Zn = (165° + 165° + 166° + 167°)/4 = 165.75° = 166°$.

Intercept − $T = (6' + 5' + 7' + 6')/4 = 6$ nautical miles.

I am confident that the plot of the average of these sights enhances the accuracy of the plotted LOP.

Combining the averaged sight with several earlier or later sights that have been averaged in the same manner yields a very good determination of a terrestrial position. I recommend this method when greater accuracy is needed for passing obstructions and making landfalls.

FINDING THE WIND

I have found the following technique useful for finding more wind when sailing in the Northeast Tradewinds. Use an accurate barometer indicating small changes in atmospheric pressure. If the winds are light and the barometer begins to rise, alter the course several degrees to the south. The higher barometric pressure indicates that the high pressure area is bending, or moving southward. The winds in the high pressure area are lighter and hence, the area is avoided. If more wind is encountered than needed, sail the great circle course to the destination.

I would like to have a recording type barometer. The recording barometer tracks pressure changes with respect to time. The recording ability of the device makes it easier to study the relationships of pressure versus wind changes. Also, a recording type barometer is more useful than the non-recording type for personal weather forecasting.

SIGHTING AND IDENTIFYING AN UNKNOWN STAR

At times I want to be able to take a sight on an unknown star. Suppose the sky has been overcast for some time one evening, and during the twilight period, a single body is observed through a hole in the cloud cover. I decide to take a sight on the unknown celestial body.

The method for taking a star or planet sight with an unknown altitude is to set the sextant at approximately 0°, turn the instrument upside down, and grasp the handle with the left hand. Observe the body through the telescope, depress the release lever with the right hand, and move the index arm forward until the horizon comes into view. Turn the sextant right side up, observe the body near the horizon, rotate the micrometer drum until the body touches the horizon. Time the sight in the usual manner.

The altitude, time, and approximate azimuth are recorded. The approximate azimuth is obtained by observing the direction toward the body from the binnacle or a hand-bearing compass and adding or subtracting the variation to obtain an approximation of the true azimuth.

The sextant altitude is corrected for index error and dip to obtain the apparent altitude. This is corrected for the remaining altitude corrections to obtain the observed altitude.

The time and assumed longitude are used to calculate the LHA of Aries.

Take out the star finder and select the blue grid template closest to the assumed latitude. Be sure that the appropriate side of the grid and base, north or south, are used and correspond to the hemisphere of the observer. Place the grid over the pin on the base and rotate the grid until the arrow on the grid is pointing to the calculated LHA of Aries.

Look up the altitude and azimuth coordinates on the blue grid. If the body sighted is one of the selected stars printed on the base, the star is identified within close proximity of the coordinates on the base.

If the star is not one of the selected stars on the base, there is another step toward identification. With the arrow on the blue template positioned on the LHA of Aries, place the red template, same hemisphere, north or south, over the blue template. Rotate the red template until the slotted meridian intersects with the altitude and azimuth coordinates of the unknown body on the blue grid.

The declination is read from the scale of the slotted meridian. If the reading is greater than ±30°, the declination is read beyond the slot. 0° declination is in the middle of the slot. The solid lines inside of 0° declination are for measuring declinations of the same name as latitude. The broken lines outside of the 0° declination are for measuring declinations of contrary name to latitude.

The quantity 360° − SHA is obtained from the base plate underneath the red template arrow and the quantity obtained is subtracted from 360° to obtain the SHA of the body. With the approximate SHA and declination known, go to the tables headed: STARS, 19— JANUARY–JUNE or STARS, 19— JULY–DECEMBER in the white pages of your current-year almanac. Contained in these tables is a list of over 170 stars listed in order of increasing SHA. Several stars close to the determined SHA can be found. However, when the declination is compared, the unknown star is easily identified. Knowing the exact value of the SHA and declination, the sight is reduced and the LOP plotted in the usual manner. If the body cannot be found in the star lists, it is probably a planet.

As an example, use the 1986 *Nautical Almamac* or Figure 9.7 to calculate the LHA of Aries as follows:

Date: 9/21/86; D. R. Long.: 150°40′ W; D. R. Lat.: 23°55′ N; Ho: 44°20′, Zn: 170°; GMT: 4h 38m 0s; and the celestial body sighted appeared to be relatively dim.

9/22/86 4h	60°43.6′
38m 0s	9°41.1′
	360°
Total GHA	430°24.7′
Ass. Long.	150°24.7′
LHA of Aries	280°

Place the blue grid labelled: LATITUDE 25N on the coordinates: altitude 44° and azimuth 170° at LHA 280°. Note there is no star in the vicinity of these coordinates. Place the red template side labelled: NORTH LAT. over the pin, and the slot on the coordinates, obtaining: Dec. S 21° and 360° − SHA = 288°. Therefore, SHA = 72°.

Open the almanac to the white page headed: STARS, 1986 JULY–DECEMBER or Figure 13.1.

Go down the first set of columns, in bold print headed: S.H.A. to 72° and find: π Sagittarii.

Go across to the second set of columns, in bold print headed: Declination and find: S. 21°.

Looking a few degrees up or down the columns, note there are other stars close to π Sagittarii in SHA and Dec. However, π Sagittarii is the closest selection for SHA and Dec.

Go to the columns headed: SEPT. Obtain and record: SHA: 72°45.4′ and Dec.: 21°02.9′ S. With the above information, the sight is reduced and plotted in the usual manner.

STARS, 1986 JULY—DECEMBER 269

Mag.	Name and Number		S.H.A.							Declination						
				JULY	AUG.	SEPT.	OCT.	NOV.	DEC.		JULY	AUG.	SEPT.	OCT.	NOV.	DEC.
			°	′	′	′	′	′	′	°	′	′	′	′	′	′
3·4	γ Cephei		**5**	18·2	17·6	17·4	17·5	18·0	18·6	**N. 77**	33·1	33·3	33·5	33·6	33·8	33·9
2·6	*Markab*	57	**13**	58·6	58·4	58·4	58·4	58·4	58·5	**N. 15**	07·9	08·0	08·1	08·1	08·1	08·1
2·6	*Scheat*		**14**	13·2	13·0	12·9	12·9	13·0	13·1	**N. 28**	00·4	00·6	00·7	00·8	00·8	00·8
1·3	*Fomalhaut*	56	**15**	46·2	46·0	45·9	45·9	46·0	46·1	**S. 29**	41·5	41·5	41·6	41·6	41·7	41·7
2·2	β Gruis		**19**	31·7	31·4	31·4	31·4	31·5	31·7	**S. 46**	57·2	57·2	57·3	57·4	57·5	57·5
2·9	α Tucanæ		**25**	35·6	35·4	35·3	35·4	35·6	35·9	**S. 60**	19·5	19·6	19·7	19·8	19·9	19·9
2·2	*Al Na'ir*	55	**28**	08·8	08·6	08·6	08·6	08·8	08·9	**S. 47**	01·5	01·6	01·6	01·7	01·8	01·8
3·0	δ Capricorni		**33**	25·3	25·2	25·2	25·2	25·3	25·4	**S. 16**	11·3	11·3	11·3	11·3	11·4	11·4
2·5	*Enif*	54	**34**	07·0	06·9	06·8	06·9	07·0	07·1	**N. 9**	48·7	48·8	48·9	48·9	48·9	48·9
3·1	β Aquarii		**37**	17·1	16·9	16·9	17·0	17·1	17·2	**S. 5**	37·9	37·8	37·8	37·8	37·8	37·9
2·6	*Alderamin*		**40**	25·7	25·7	25·8	26·0	26·3	26·6	**N. 62**	31·5	31·7	31·8	31·9	32·0	32·0
2·6	ε Cygni		**48**	34·8	34·7	34·8	34·9	35·1	35·2	**N. 33**	55·0	55·2	55·3	55·4	55·4	55·3
1·3	*Deneb*	53	**49**	45·1	45·1	45·2	45·3	45·5	45·7	**N. 45**	13·7	13·9	14·1	14·1	14·1	14·1
3·2	α Indi		**50**	50·2	50·1	50·2	50·3	50·5	50·6	**S. 47**	20·4	20·4	20·5	20·6	20·6	20·6
2·1	*Peacock*	52	**53**	50·6	50·5	50·6	50·8	51·0	51·2	**S. 56**	46·8	46·9	47·0	47·0	47·0	47·0
2·3	γ Cygni		**54**	33·6	33·5	33·6	33·8	34·0	34·1	**N. 40**	12·6	12·8	12·9	13·0	13·0	12·9
0·9	*Altair*	51	**62**	27·9	27·9	27·9	28·0	28·1	28·2	**N. 8**	49·8	49·9	50·0	50·0	50·0	49·9
2·8	γ Aquilæ		**63**	35·5	35·5	35·5	35·6	35·8	35·8	**N. 10**	34·7	34·8	34·8	34·9	34·8	34·8
3·0	δ Cygni		**63**	51·4	51·4	51·5	51·7	51·9	52·0	**N. 45**	05·7	05·9	06·0	06·0	06·0	05·9
3·2	*Albireo*		**67**	27·1	27·1	27·2	27·3	27·5	27·5	**N. 27**	55·7	55·9	56·0	56·0	55·9	55·9
3·0	π Sagittarii		**72**	45·3	45·3	45·4	45·5	45·6	45·6	**S. 21**	02·9	02·9	02·9	02·9	02·9	02·9
3·0	ζ Aquilæ		**73**	47·9	47·9	48·0	48·1	48·2	48·3	**N. 13**	50·5	50·6	50·6	50·6	50·6	50·5
2·7	ζ Sagittarii		**74**	33·4	33·4	33·5	33·6	33·7	33·7	**S. 29**	54·1	54·2	54·2	54·2	54·2	54·2
2·1	*Nunki*	50	**76**	23·3	23·3	23·4	23·5	23·6	23·6	**S. 26**	19·0	19·0	19·0	19·0	19·0	19·0
0·1	*Vega*	49	**80**	52·4	52·5	52·7	52·8	53·0	53·1	**N. 38**	46·2	46·3	46·4	46·4	46·3	46·2
2·9	λ Sagittarii		**83**	12·7	12·7	12·8	12·9	13·0	13·0	**S. 25**	25·9	25·9	26·0	26·0	26·0	25·9
2·0	*Kaus Australis*	48	**84**	10·6	10·6	10·7	10·8	10·9	10·9	**S. 34**	23·7	23·7	23·7	23·7	23·7	23·7
2·8	δ Sagittarii		**84**	57·8	57·8	57·9	58·0	58·1	58·1	**S. 29**	50·2	50·2	50·3	50·3	50·3	50·2
3·1	γ Sagittarii		**88**	45·6	45·7	45·8	45·9	46·0	46·0	**S. 30**	25·7	25·7	25·7	25·7	25·7	25·7
2·4	*Eltanin*	47	**90**	55·2	55·3	55·5	55·8	56·0	56·1	**N. 51**	29·4	29·5	29·6	29·6	29·5	29·3
2·9	β Ophiuchi		**94**	17·7	17·7	17·9	18·0	18·1	18·0	**N. 4**	34·3	34·3	34·3	34·3	34·3	34·2
2·5	κ Scorpii		**94**	36·4	36·5	36·6	36·8	36·9	36·8	**S. 39**	01·6	01·7	01·7	01·7	01·6	01·6
2·0	θ Scorpii		**95**	54·5	54·6	54·7	54·9	55·0	54·9	**S. 42**	59·6	59·7	59·7	59·7	59·6	59·5
2·1	*Rasalhague*	46	**96**	25·2	25·2	25·3	25·5	25·5	25·5	**N. 12**	34·1	34·2	34·2	34·2	34·1	34·0
1·7	*Shaula*	45	**96**	49·4	49·4	49·6	49·7	49·8	49·7	**S. 37**	05·9	05·9	05·9	05·9	05·9	05·8
3·0	α Aræ		**97**	17·7	17·8	18·0	18·2	18·3	18·2	**S. 49**	52·2	52·3	52·3	52·3	52·2	52·1
3·0	β Draconis		**97**	27·6	27·8	28·1	28·3	28·5	28·5	**N. 52**	18·7	18·8	18·8	18·8	18·7	18·5
2·8	υ Scorpii		**97**	32·1	32·1	32·3	32·4	32·5	32·4	**S. 37**	17·3	17·4	17·4	17·4	17·3	17·3
2·8	β Aræ		**98**	57·0	57·1	57·3	57·5	57·6	57·6	**S. 55**	31·3	31·4	31·4	31·4	31·3	31·2
Var.‡	α Herculis		**101**	29·3	29·4	29·5	29·6	29·7	29·7	**N. 14**	24·3	24·3	24·4	24·3	24·3	24·2
2·6	*Sabik*	44	**102**	35·8	35·8	35·9	36·0	36·1	36·0	**S. 15**	42·7	42·6	42·6	42·6	42·6	42·7
3·1	ζ Aræ		**105**	37·2	37·3	37·5	37·8	37·8	37·8	**S. 55**	58·4	58·5	58·5	58·5	58·4	58·3
2·4	ε Scorpii		**107**	40·5	40·6	40·7	40·8	40·9	40·8	**S. 34**	16·4	16·4	16·4	16·4	16·3	16·3
1·9	*Atria*	43	**108**	11·1	11·4	11·7	12·1	12·2	12·1	**S. 69**	00·6	00·6	00·6	00·6	00·5	00·4
3·0	ζ Herculis		**109**	48·1	48·2	48·3	48·5	48·6	48·5	**N. 31**	37·7	37·7	37·7	37·7	37·6	37·4
2·7	ζ Ophiuchi		**110**	53·6	53·7	53·8	53·9	53·9	53·9	**S. 10**	32·5	32·5	32·5	32·5	32·5	32·6
2·9	τ Scorpii		**111**	14·2	14·3	14·4	14·6	14·6	14·5	**S. 28**	11·5	11·5	11·5	11·5	11·5	11·4
2·8	β Herculis		**112**	35·2	35·3	35·4	35·6	35·6	35·5	**N. 21**	31·1	31·2	31·2	31·1	31·0	30·9
1·2	*Antares*	42	**112**	51·2	51·3	51·4	51·5	51·5	51·4	**S. 26**	24·3	24·3	24·3	24·3	24·3	24·3
2·9	η Draconis		**114**	02·4	02·6	03·0	03·3	03·5	03·4	**N. 61**	32·8	32·9	32·9	32·8	32·6	32·4
3·0	δ Ophiuchi		**116**	35·3	35·4	35·5	35·6	35·6	35·5	**S. 3**	39·7	39·7	39·6	39·7	39·7	39·8
2·8	β Scorpii		**118**	50·1	50·2	50·3	50·4	50·4	50·3	**S. 19**	46·3	46·3	46·3	46·3	46·2	46·3
2·5	*Dschubba*		**120**	06·9	07·0	07·1	07·2	07·2	07·1	**S. 22**	35·2	35·2	35·2	35·2	35·1	35·1
3·0	π Scorpii		**120**	29·4	29·5	29·6	29·7	29·7	29·6	**S. 26**	04·7	04·7	04·7	04·7	04·7	04·7
3·0	β Trianguli Aust.		**121**	30·7	30·9	31·2	31·4	31·4	31·2	**S. 63**	23·7	23·8	23·7	23·7	23·5	23·4
2·8	α Serpentis		**124**	05·9	05·9	06·0	06·1	06·1	06·0	**N. 6**	28·0	28·1	28·1	28·0	28·0	27·9
3·0	γ Lupi		**126**	26·4	26·5	26·6	26·7	26·7	26·6	**S. 41**	07·6	07·6	07·5	07·5	07·4	07·4
2·3	*Alphecca*	41	**126**	28·1	28·2	28·3	28·4	28·5	28·4	**N. 26**	45·7	45·7	45·7	45·6	45·5	45·3

‡ 3·0—3·7

FIGURE 13.1 ☆ *NAUTICAL ALMANAC*, STARS PAGE, JULY—DECEMBER (TYPICAL)

SIGHTING AND IDENTIFYING AN UNKNOWN PLANET

The procedure for sighting and identifying an unknown planet is essentially the same as for sighting and identifying an unknown star up to and including the determination of the Dec., and 360° − SHA. One clue is that the Dec. of a planet is always less than 30° north or south. The SHA is added to the GHA of Aries, obtaining the approximate GHA of the planet. With GHA and Dec., the *Nautical Almanac* is consulted. The G.H.A. and Dec. columns under VENUS, MARS, JUPITER, and SATURN are inspected at the time of the sight to identify the planet. With exact GHA and Dec., the sight is reduced, and the LOP plotted.

Follow the example using the 1986 *Nautical Almanac*, or Figure 9.7, for calculating the LHA of Aries, and for finding the planet:

Date: 9/21/86; D. R. Long.: 151°04′ W; D. R. Lat.: 23°42′ N; Ho: 55°10′; Zn: 150°; GMT: 8h 3m 55s on 9/22/86; and the celestial body appeared to be very bright:

9/22/86 8hr	120°53.4′
3m 55s	0°58.9′
	360°
Total GHA	481°52.3′
Ass. Long.	150°52.3′W
LHA of Aries	331°

Place the blue grid for LATITUDE 25N on the coordinates: altitude 55° and azimuth 150° at LHA 331° and the red template side labelled: NORTH LAT. over the pin, and the slot on the coordinates, obtaining: Dec. 7° S and 360° − SHA = 348°. Therefore, SHA = 12°.

The only stars close to the SHA of 12° are Markab and Scheat. However, their Dec.'s are not close. Because of the declination and brilliance of the celestial body, I suspect it to be a planet.

Add the GHA of Aries: 121°52.3′ to the SHA: 12° and obtain the approximate GHA: 133°52.3′.

Inspect the planet columns at the GMT of the sight. Using the almanac or Figure 9.7, note that for Jupiter the GHA and Dec. at

8 hrs are 132°50.6′ and 6°46.6′ S respectively, indicating that Jupiter has been sighted.

Sight reduction and plotting are accomplished in the usual manner.

DETERMINING COMPASS ERROR AT SEA

One of the navigator's responsibilities is to check the *compass error* underway. Compass error is the difference between true direction and compass heading.

This task is easily accomplished in the harbor by attaching an azimuth circle device to the compass. The azimuth circle is adjusted to be level, and the reflection from the sun is focused into a thin line of light that is easily reflected for reading on the compass card. The circle is adjusted to the timed azimuth of the sun, and the magnetic variation is either added or subtracted to determine the deviation of the compass for the heading. This procedure is repeated for a number of timed azimuths and headings around the compass. The resulting deviations are recorded, and a card is prepared showing deviations of any number of headings, usually at about fifteen-degree intervals.

On a small craft at sea there are difficulties in obtaining the deviation of the compass because the only level surface is the compass card. There is a shadow that is created by the rays of the sun passing over the *shadow pin* in the center of the card. Therefore, it is necessary to observe the compass reading where this shadow falls on the card.

It is important to check the compass early in the morning or late in the afternoon. The times are critical because low altitude azimuths are required for this procedure. It will take some experience to determine the optimum times of low altitude of the sun, early morning or late afternoon, to obtain the readings on the compass card. Do not attempt to swing or make any adjustments of the compass at sea. We are only concerned with determining the deviation, not changing it.

For the following example, open the 1986 *Nautical Almanac* to the the right hand white page headed: 1986 SEPTEMBER 22, 23, 24 (MON., TUES., WED.) or go to Figure 9.4.

At about 0600 hours local time on 9/22/86 we decide to calculate the sun's azimuth for 0700 hours, or 1700 hours GMT. The approximate DR Lat. and DR Long. will be 23°30′ N and 151°45′ W respectively.

9/22/86 17h	76°49.3′	Dec. 0°14.6′ N
	360°	
Total GHA	436°49.3′	
Ass. Long.	151°49.3′	
LHA	285°	

Open the almanac to the page headed: LATITUDE/A: 18°–23° and use the above information just as you would to reduce a sight. You will find that Z = Zn = 95°. Note that Hc = 13°53′. The altitude is sufficiently low and the sun will cast a good shadow from the pin in the center of the compass card to the graduations on the opposite side of the card.

Tune your receiver to stations WWV and WWVH and at exactly 1700 hours GMT take a reading from the narrow shadow band on the opposite side of the card. The reading is a little shy of the 264° graduation and appears to be approximately 263°. The magnetic variation in this area of the North Pacific Ocean, 12°, is added to 263°, obtaining 275°. Calculate the reciprocal of this number by subtracting 180° and obtain 95°, the *shadow bearing*. The azimuth was 95°, the same as the shadow bearing, so there is no deviation for the heading. If deviation is found, it should be recorded and the heading changed to compensate for it. The normal procedure on a yacht is to check the deviation every few days or when the heading is changed.

OBSERVATION ERRORS

Basic celestial navigation is not complete without some discussion of observation errors. These errors are reduced to three categories as follows:

1. *Instrument error*—Constant error built into a sextant and index error. In sight reduction, instrument error is systematically eliminated by the application of corrections.
2. *Personal error*—Error incurred from observations of a specific observer, e.g., the tendency of an observer to over- or underestimate the horizon. The error is usually discovered eventually, and a correction applied for the magnitude of the over- or underestimate.
3. *Random error*—Non-systematic error due to unpredictable circumstances. Such variable errors average to zero, if constant-type errors like instrument and personal errors are compensated for.

With practice, observation errors can be eliminated.

ERRORS IN PLOTTING FIXES

The least error range occurs when two sights used to determine a terrestrial position are at right angles (90°) to each other. A greater error range is introduced by any angle other than 90° between the azimuths as shown in Figure 13.2. The most probable position is somewhere within the zone defined by the margin of error.

The first illustration in Figure 13.3 is of a proper three-star fix with azimuths differing by 120°. Suppose there was a systematic error adding some positive amount to the altitudes of the sights. This error would expand the size of the equilateral triangle defining the margin of the error zone. However, since the expanded zone is equally distant from the center of the fix in all directions, the most probable position of the observer is still at the center of the triangle.

The second illustration in Figure 13.3 is of a three-star fix with azimuths differing by only 60° and with the same positive error as described above. The triangle formed by the 60° azimuths would appear to be equal to the triangle formed by the 120° azimuths, but the similarity ends there. This is because in the second example the error zone displaced in the same direction of the azimuths would displace the most probable position of the observer away from the center of the fix.

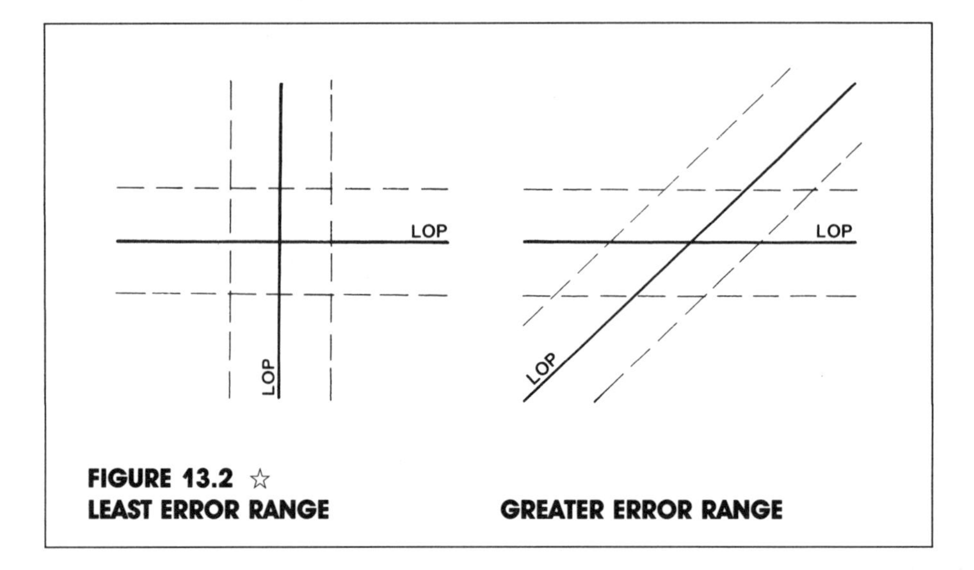

FIGURE 13.2 ☆
LEAST ERROR RANGE **GREATER ERROR RANGE**

I will leave it as an exercise for you to draw the two illustrations with a negative systematic error and show that with 120° azimuths the observer's position would remain at the center of the triangle, and with the same 60° azimuths the observer's most probable position would be displaced to the opposite side of the triangle.

I hope you are convinced that the proper separation of azimuths for a three-star fix would be to select stars as close to 120° as can be obtained.

When three observations are made in adverse weather and sea conditions, the resulting plot of the LOPs may yield a large triangle. If it is possible to obtain a fourth sight, one of the sights may be rejected, resolving the dilemma. However, if the fourth sight does not resolve the problem, and if the three observations were made within an azmuth range of 180° or less, it is possible that the most probable position is outside of the triangle. The most probable position is determined by bisecting the azimuths and plotting the bisectors, as shown in Figure 13.4.

The three azimuths are as follows: Z1 = 112°, Z2 = 152°, and Z3 = 177°, all within 180° of each other. Their bisectors are as follows: (Z1 + Z2)/2 = (112° + 152°)/2 = 132°; (Z2 + Z3)/2 = (152° + 177°)/2 = 144.5°; and (Z1 + Z3)/2 = (112° + 177°)/2 = 164.5°.

The bisectors are plotted and the most probable terrestrial position is displaced to a position outside of the triangle.

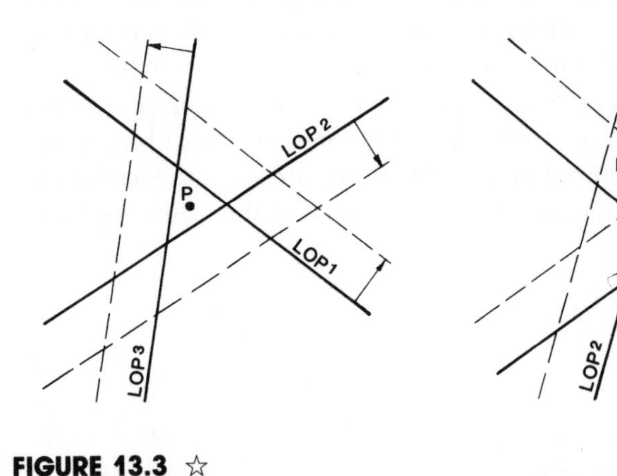

FIGURE 13.3 ☆
120° AZIMUTH SEPARATION
Z1 = 30°, Z2 = 150°,
& Z3 = 270°

60° AZIMUTH SEPARATION
Z1 = 30°, Z2 = 270°,
& Z3 = 330°

Basic celestial navigation has been covered at length in the preceding chapters. If you understand the material contained herein, can apply it effectively offshore, have a well-found yacht, and possess the necessary sailing skills, you can go cruising anywhere you would like to go.

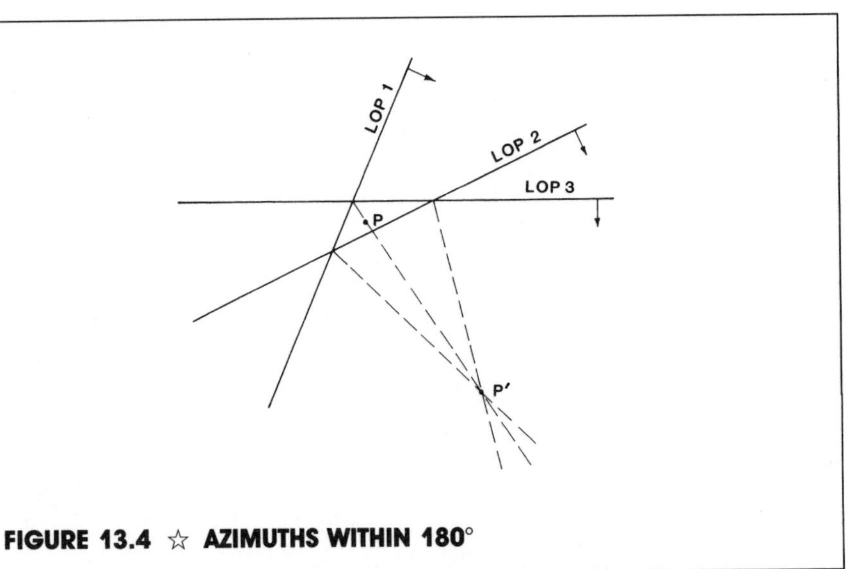

FIGURE 13.4 ☆ AZIMUTHS WITHIN 180°

Sooner or later you will want to enhance your navigational skills. When this occurs, continue on to the remaining theory and calculator-navigation chapters. These constitute an intermediate treatment of celestial navigation. The material is more comprehensive, and is designed to more than adequately prepare a world-navigator. It is presented to stimulate those truly interested in celestial navigation, and help navigators to begin collecting their own repertoire of useful techniques.

BOOK 2

INTERMEDIATE CELESTIAL BY COMPUTING METHODS

ABOARD 'ZOE'

FOURTEEN

THE THEORY

I purposely left the theory for last. The theory is not essential for the successful practice of celestial navigation using the sight reduction tables. However, if you want to become a better navigator, and not be subject to the limitations inherent in tabular solutions, you should learn the theory and then some intermediate methods utilizing a scientific calculator.

Let's begin the discussion by considering a reasonable way to establish the relationship of the celestial bodies with respect to the earth. For the purpose of celestial navigation it is useful to imagine that the earth is a true sphere, located at the center of the universe, with all the celestial bodies rotating around an extension of the earth's *polar axis*. The sun, moon, stars, and planets are imagined to be located on the transparent surface of a sphere of infinite radius, called the *celestial sphere*. Because all the bodies are located at an infinite distance from the earth, the light rays from any one of these bodies would arrive at any point on the earth's surface parallel to each other.

Let's discuss this last statement in detail. Suppose that you are standing near the base of a steel tower and the tower is supported by four equal-length cables extending from the top of the tower to a level pad, as shown in Figure 14.1. Think of the cables as light rays emanating from a pin point light source, like a star, at the top of the tower. If you measured the angles that the cables make with the level pad, you find that all four angles are equal to each other and less than 90°. The tower is vertical and therefore at a 90° angle with the pad. Also notice, the distance from the base of the tower to the cable mounts on the pad are identical for all four cables.

Now think of the base of the tower as a point on the earth's surface directly in line with the light source, or body, on top of the tower and extending through the surface of the earth to a point exactly at the earth's center. The point at the earth's surface is called the *geographical position* (GP). A light ray emanating from a body to the earth at the GP of the body is always equal to 90° with the earth's surface.

Now suppose the steel tower is extended to twice its length and the steel cables are proportionately extended without changing the distance from their mounts on the pad to the base of the tower. The angle that any one of the steel cables makes with the pad increases, but is still less than 90°.

If the steel tower is extended an *infinite* distance from the earth onto the transparent surface of the celestial sphere, the cables are extended proportionately, and the cable mounting distance from the pad to the base of the tower is not changed, the angle between the pad and the cables, for practical purposes, becomes 90°.

Since the pad is flat, the cables could be moved any finite distance, like the radius of the earth, from the tower and still be at 90°. Therefore, if the rays of light from a body on the transparent surface of the celestial sphere are thought of in the same manner as the steel cables, the light rays reach the earth's surface parallel to each other.

Let's change the flat surface of the pad to the curved surface of the earth, and see how this influences the arrival of parallel light rays from a celestial body. See Figure 14.2.

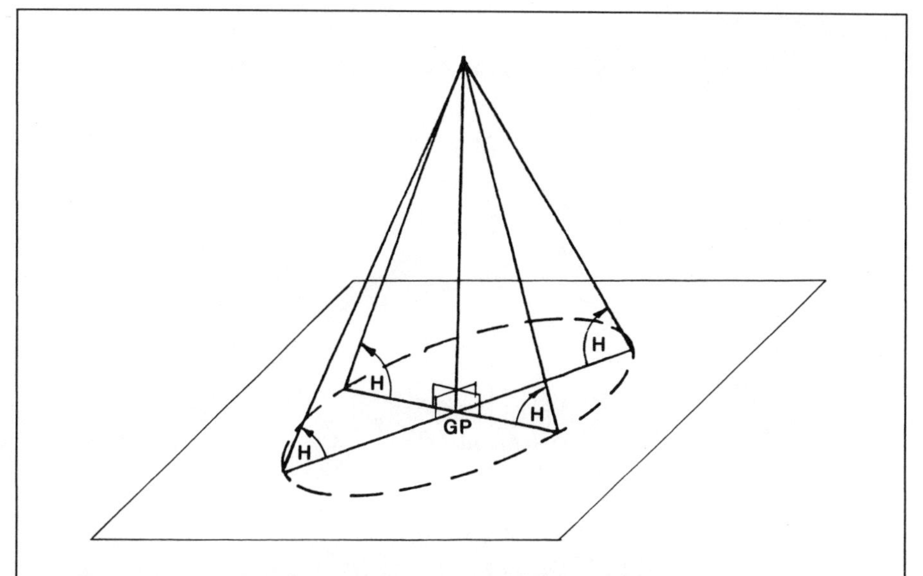

FIGURE 14.1 ☆ STEEL TOWER SUPPORTED BY FOUR EQUAL CABLES

At the GP the light rays are always at 90° with the earth's surface. At any other position on the earth's surface the light rays arrive at some angle less than 90°. The angle described is the observed altitude (Ho) that you are familiar with.

Suppose Ho was 75° with respect to the earth's surface, or the *horizon*. A circle could be drawn around the GP on the earth's surface connecting all points where Ho is 75°. The circle describes a circle of position on the earth's surface around the geographical position, and is called a *circle of equal altitude*.

Now, suppose we remove a small segment of the circle of position, or equal altitude, very near to some point on the earth's surface that has been designated as an assumed position (AP). The segment of the circle is so small that it can be redrawn as a straight line. Hence, the segment of the circle of position can be redefined as a line of position (LOP). Therefore, a circle of position contains an infinite number of lines of position for an infinite number of assumed positions.

Suppose a second body is observed. A second circle of position can be generated on the earth's surface. Visualize the intersection of the two circles of position. There are only two possible intersections of the two circles. The observer is at one of the intersec-

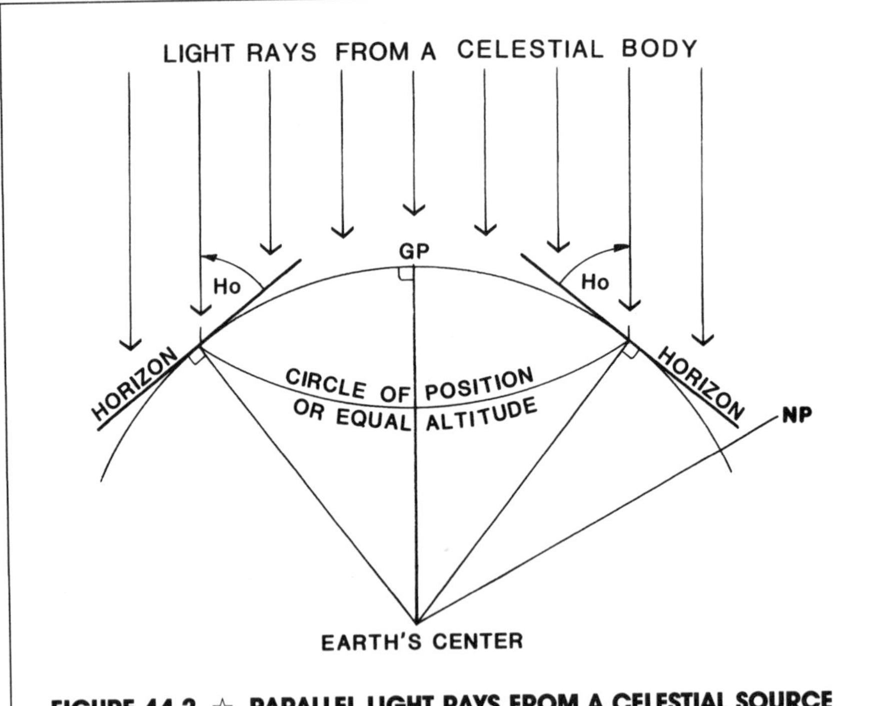

FIGURE 14.2 ☆ PARALLEL LIGHT RAYS FROM A CELESTIAL SOURCE

tions. In reality, two intersections of circles of position are separated by large geographical distances. Therefore, the navigator is not confused in making an appropriate selection of the intersection representing the terrestrial position, or position fix. The selection of a position very near to this intersection is no more than the assumed position concept we derived from the DR position earlier.

Let's return to the celestial sphere. The celestial sphere is infinitely large because it has an infinite radius. The concept of the infinite sphere helps us to see how the rays of light from a body arrive parallel to each other at the earth's surface and how circles of position are developed into lines of position.

Lines of latitude and longitude are called *cartesian coordinates.* These coordinates, which define an observer's position on the earth's surface, are defined in reference to the equator and the *prime* or *Greenwich meridian*, respectively. A similar coordinate system is extended into the universe, to the transparent surface of the celestial sphere, as shown in Figure 14.3. You have been using this set of

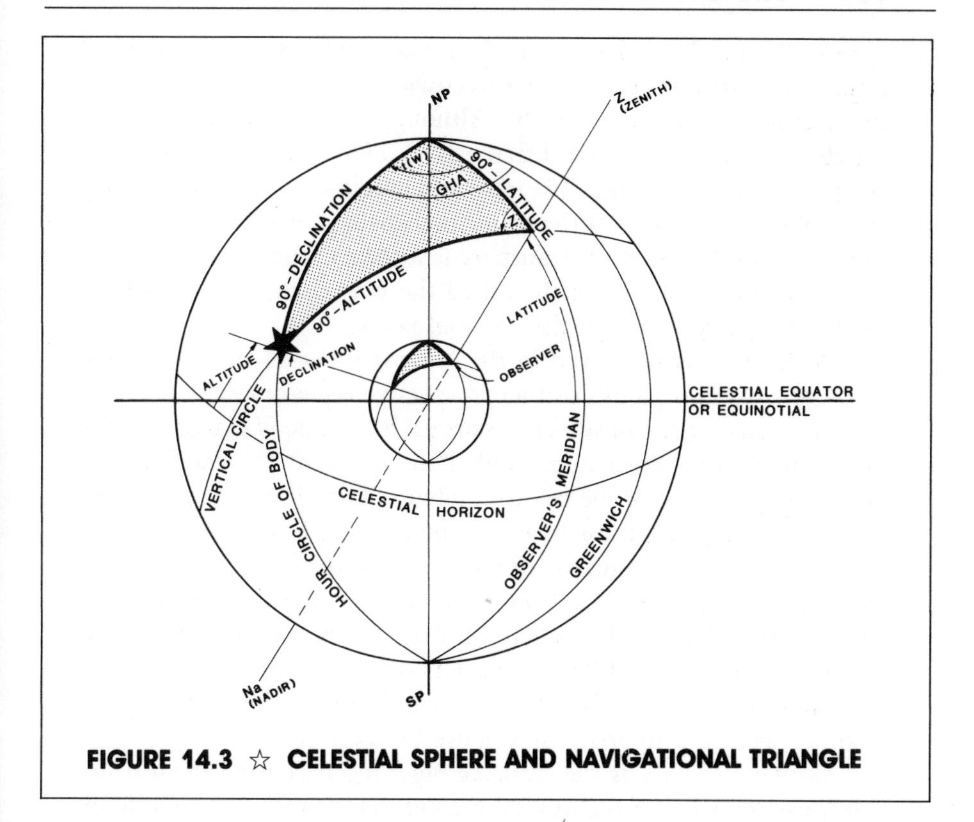

FIGURE 14.3 ☆ CELESTIAL SPHERE AND NAVIGATIONAL TRIANGLE

coordinates, called the *celestial coordinates*, to define the positions of celestial bodies.

When the polar axis of the earth is extended outward, north and south to infinity, the axis forms the north and south polar axis of the celestial sphere (NP) & (SP), called the *celestial poles*.

The plane of the earth's equator is extended into space forming a *great circle* called the *celestial equator*, or *equinoctial*, where the equatorial plane intersects the transparent surface of the celestial sphere. The celestial poles and the equinoctial form the reference system called the *polar-equitorial system* of coordinates.

The prime meridian, or *Greenwich meridian*, is similarly extended to the transparent surface of the celestial sphere. The position of the prime meridian is defined to be zero degrees in space, as well as on earth, and extends from the north polar axis to the south polar axis of the celestial sphere and is therefore a great circle.

A body, such as a star, on the transparent surface of the celestial sphere can be defined positionally by coordinates similar to longitude and latitude, called the Greenwich hour angle (GHA) and the declination (d), respectively. Although GHA is similar to longitude, GHA is measured differently. GHA is measured west of the prime meridian from 0° to 360°, and longitude is measured from 0° to 180°, east or west of the prime meridian. The GHA of a body defines a coordinate on the transparent surface of the celestial sphere similar to longitude on earth, called the *hour circle*, and is a great circle that passes through the celestial poles.

Declination is measured in the same manner as latitude, from 0° to 90°, north or south of their respective equators.

The plane of the observer's assumed longitude is extended into space, forming a great circle called the *observer's meridian*, at the transparent surface of the celestial sphere. The observer's meridian in space corresponds numerically to the assumed longitude on earth.

The local hour angle, or LHA, of a celestial body is measured westward, from 0° to 360°, from the observer's meridian to the hour circle of the body. LHA is determined numerically by adding the assumed longitude to GHA in East Longitude and subtracting assumed longitude from GHA in West Longitude. Add or subtract 360° as necessary to arrive at a figure between 0° and 360°.

Similar to LHA is the *meridian angle* (t), measured from 0° to 180°, east or west of the observer's meridian, to the hour circle of the body. "t" is labelled: E or W to indicate the direction of the measurement. For tabular solutions, LHA was convenient. However, the meridian angle (t), will be used for calculations in the remainder of this material.

Another system of coordinates must be defined in reference to the observer's horizon and not the celestial equator. The system is called the *zenith-horizon system* of coordinates. The horizon of the observer is extended outward onto the transparent surface of the celestial sphere and forms a great circle called the *celestial horizon*.

Perpendicular to the celestial horizon is the observer's *zenith* (Z). The observer's zenith is extended outward to where it intersects with the transparent surface of the celestial sphere.

The line defined by the zenith is extended below the observer, through the center of the earth, and outward to where it intersects with the transparent surface of the celestial sphere on the other side of the celestial sphere, a point called the *nadir* (Na).

A great circle, called the *vertical circle*, is generated on the transparent surface of the celestial sphere and passes through the observer's zenith, the body being observed, and the nadir.

The vertical circle, hour circle of the body, and the observer's meridian, all form the three sides of a spherical triangle on the transparent surface of the celestial sphere, called the *celestial triangle.*

Altitude is the angle between the celestial horizon and the body on the vertical circle.

In Figure 14.3, note that two sides of the celestial triangle, 90°-Dec. and 90°-Lat., are defined from the polar-equatorial system of coordinates, and the third side, 90°-Alt., is defined by the zenith-horizon system of coordinates. The zenith-horizon system is slightly more difficult to visualize. Azimuth (Z), the angle between 90°-Lat. and 90°-Alt., is measured in North Latitude east or west, and South Latitude east or west.

The celestial triangle and the terrestrial triangle are shaded to clarify some important similarities.

GHA is given in the almanac for the sun, moon, and planets. To find the GHA of a star, the GHA of the *first point of Aries* is added to the sidereal hour angle (SHA) of the star. SHA is the angle between Aries and a celestial body. The *GHA of Aries* is used because the SHA and declination of stars change slowly when compared to the sun, moon, and planets. And if the GHA of each star were calculated, there wouldn't be enough room in the cabin for anything but the extra volumes of the almanac. The first point of Aries is nothing more than a point in space on the transparent surface of the celestial sphere called the *vernal equinox* by astronomers. The vernal equinox is the position on the celestial equator that the sun intersects when it passes from south declination to north declination in the early spring. Because we are not astronomers, it is best to think of Aries as nothing more than a convenient great circle, meridian, or hour circle, on the transparent surface of the celestial sphere. Aries is measured west of Greenwich and SHA is measured west of Aries, as shown in Figure 14.4.

The development of the celestial triangle is necessary to understand the theory of Celestial Navigation. However, the *navigational triangle*, projected back to the earth's surface, has finite dimensions and is therefore of much greater use to the practical navigator.

The projected triangle formed on the earth's surface is called the terrestrial triangle. The terrestrial triangle is formed by the observer's elevated pole (NP), the assumed position of the observer (AP), and the geographical position of the body (GP), as shown in Figure 14.5.

The *elevated pole*, by convention and for convenience, is always

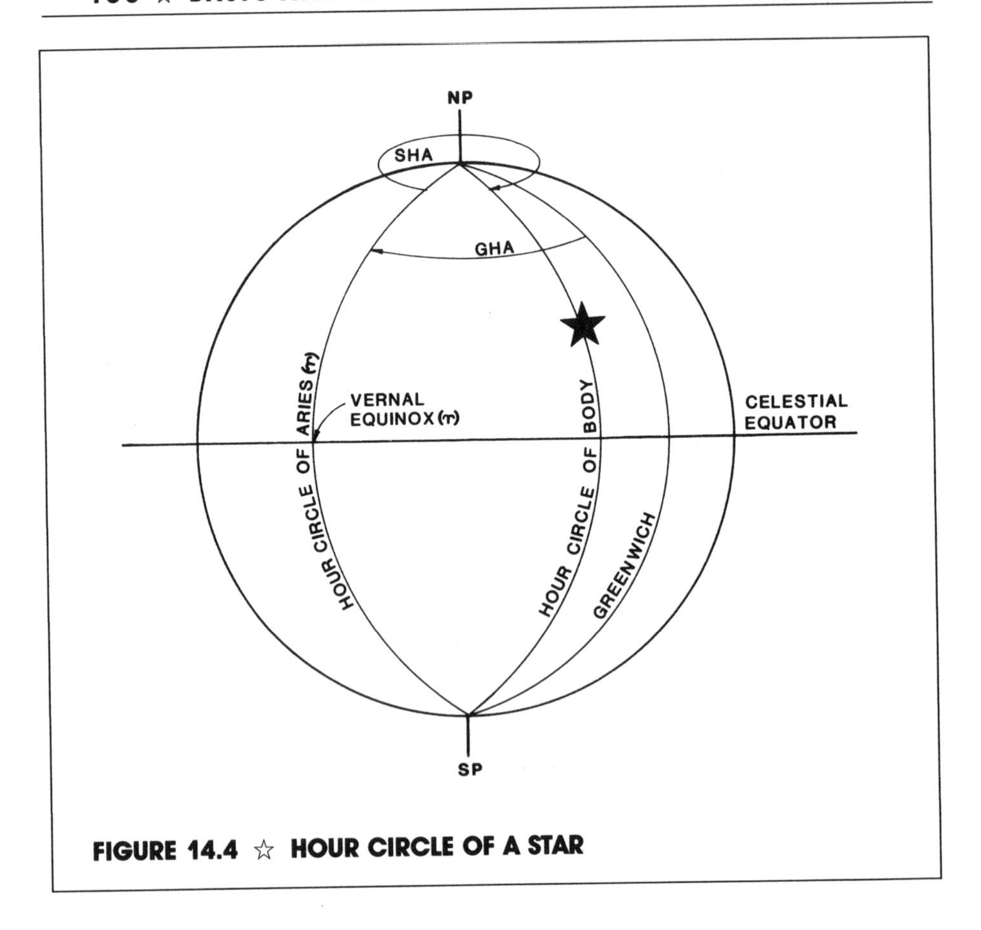

FIGURE 14.4 ☆ HOUR CIRCLE OF A STAR

of the same name as the observer's assumed latitude, north or south. The elevated pole is the pole that is above the observer's horizon.

Often an observer will take a sight of a celestial body where the body's geographical position is in the opposite hemisphere. Therefore, the geographical position of a body can be in either hemisphere regardless of the observer's latitude. The possible number of geographical and observer positions is infinite, and the navigational triangle can have an infinite variety of shapes.

It is customary to rename the sides of the navigational triangle as follows: 90°-Dec. becomes the *polar distance*, 90°-Lat. becomes the *colatitude*, and 90°-Alt. becomes the *coaltitude*.

When the declination of a body is the same name as the observer's latitude, the declination is subtracted from 90°, establishing the polar distance. When the declination of a body is a contrary

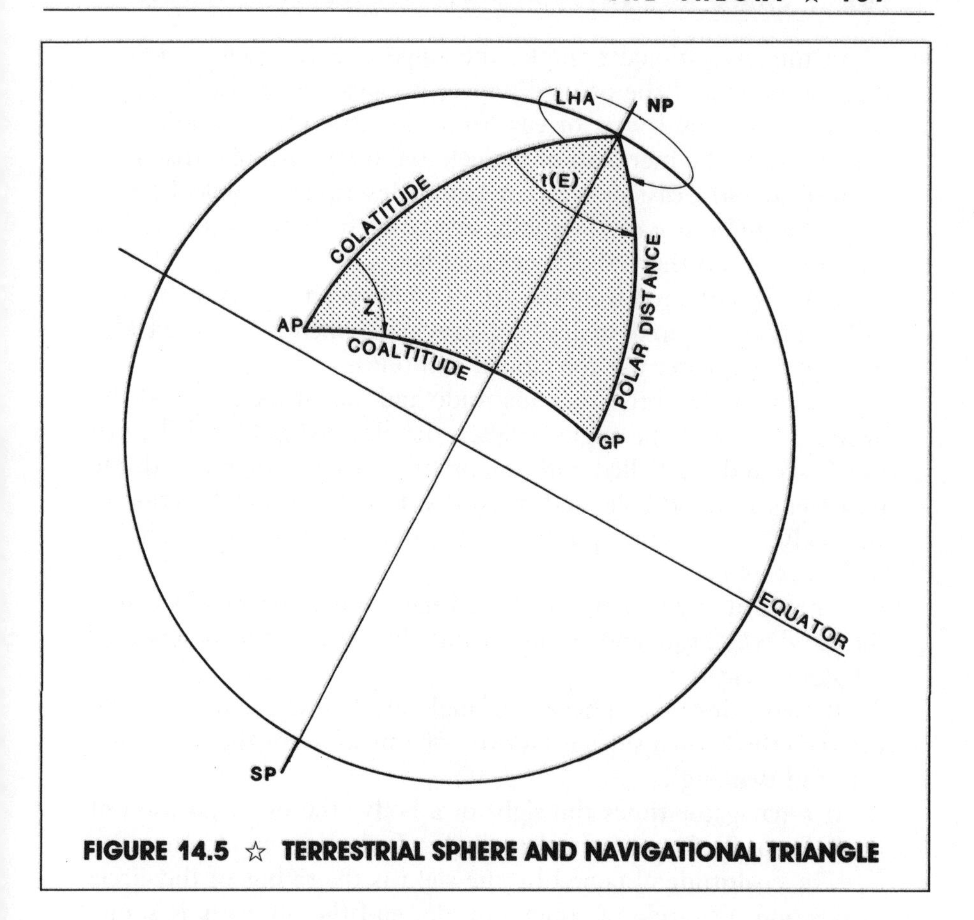

FIGURE 14.5 ☆ TERRESTRIAL SPHERE AND NAVIGATIONAL TRIANGLE

name to the observer's latitude, the declination is added to 90°, establishing the polar distance.

The assumed latitude is subtracted from 90° establishing the colatitude.

Both the polar distance and the colatitude are in the polar-equatorial system of coordinates.

The coaltitude is slightly more difficult to visualize because it is in the zenith-horizon system of coordinates. However, coaltitude is easily established by subtracting the altitude of a body from 90°.

Each side of the navigational triangle is a great circle, and the length of the side, in *nautical miles*, is calculated by converting the angular value into minutes, or *arc distance*. One minute of arc distance on the earth's surface is the same *arc measure* as one minute of angular arc at the earth's center.

In the navigational triangle, the angle at either pole, between the colatitude and the polar distance, is called the meridian angle (t). "t" is measured west or east from the observer's meridian, or colatitude, to the meridian of the celestial body, or polar distance, from 0° to 180°, east or west. A better description would be that "t" is the difference of longitude, east or west, between the AP of an observer and the GP of a celestial body.

LHA is measured from the AP of an observer, westward to the GP of a body, from 0° to 360°. As previously mentioned, the LHA concept has greater use for tabular solutions.

The angle between the colatitude and coaltitude is called the azimuth (Z). Z is the angle between the elevated pole and the GP of a body and is labelled with the prefix: N or S, to agree with the observer's elevated pole, and the suffix: E or W, in the direction of the body. For plotting purposes, Z is converted to Zn, as shown in Figure 14.6.

The third angle of the terrestrial navigational triangle is called the *parallactic angle* and is not ordinarily used in the practice of celestial navigation.

If two sides of a spherical triangle are known, and the angle between the known sides is known, we can solve for the remaining side and two angles.

If a navigator times the sight of a body, the exact position of the GP can be found from data easily obtained in an almanac.

The coaltitude obtained by the sight is the radius of the circle of position, or circle of equal altitude, and the observer is somewhere on the circumference of the circle of position.

Unfortunately, the observer's position cannot be determined with a single observation, because the azimuth of the GP, at the moment the sight was taken, cannot be determined with sufficient precision. A small segment of the circle of position, called an LOP, is taken near the AP of the observer.

With an AP selected and the GP of the body at the time of the sight known, the coaltitude, polar distance, and meridian angle (t) are calculated.

Having accomplished the above, sight reduction becomes a matter of solving two classical equations from *spherical trigonometry*. The solutions for the computed altitude (Hc) and the azimuth (Z) are found by applying the *law of cosines* from spherical trigonometry. The law states that the cosine of any side of a spherical triangle is equal to the product of the cosines of the other two sides, plus the

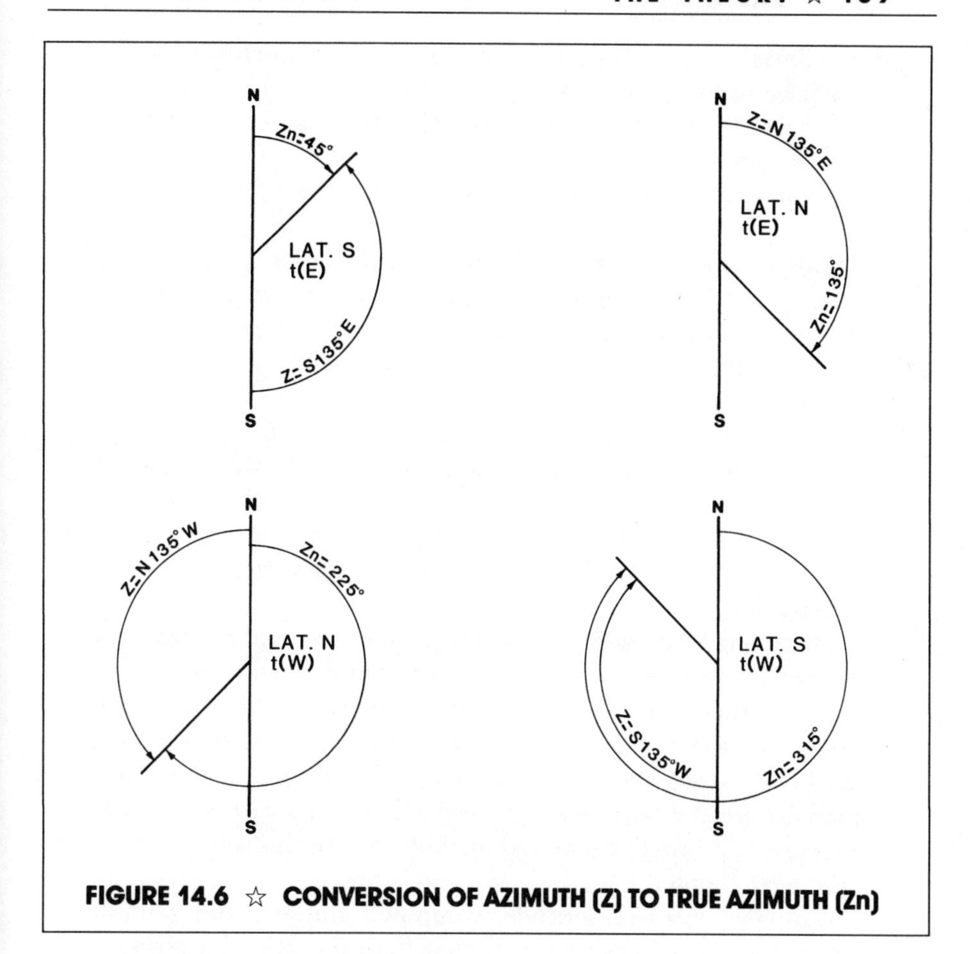

FIGURE 14.6 ☆ CONVERSION OF AZIMUTH (Z) TO TRUE AZIMUTH (Zn)

product of the sines of the other two sides and the cosine of their included angle. The law of cosines is mathematically stated for the sides, (90° − Hc), and (90° − Z), of the navigational triangle as follows:

$$\cos(90° - Hc) = \cos(90° - L)\cos(90° - d) + \sin(90° - L)\sin(90° - d)\cos t$$ and

$$\cos(90° - d) = \cos(90° - L)\cos(90° - Hc) + \sin(90° - L)\sin(90° - Hc)\cos Z$$

Two trigonometric reduction formulas are used to simplify the above formulas. The reduction formulas mathematically state that the cosine of the co-angle (90° minus the angle) is equal to the sine of the angle, and conversely, the sine of the co-angle (90° minus the angle) is equal to the cosine of the angle. The reduction formulas we will be utilizing are as follows:

$\cos(90° - Hc) = \sin Hc$, $\cos(90° - L) = \sin L$, $\cos(90° - d) = \sin d$

$\sin(90° - Hc) = \cos Hc$, $\sin(90° - L) = \cos L$, $\sin(90° - d) = \cos d$

Simpler, more functional formulas, eliminating excessive calculations, are derived by substituting the above reductions into the initial equations, which then become:

$$\sin Hc = \sin L \sin d + \cos L \cos d \cos t$$

and

$$\sin d = \sin L \sin Hc + \cos L \cos Hc \cos Z$$

Solving for cos Z, in the second formula, we obtain the following:

$$\cos Z = (\sin d - \sin L \sin Hc)/\cos L \cos Hc$$

where L, d, and t are latitude, declination and meridian angle, respectively.

These formulas, with some modification, are used to create the sight reduction tables you have become familiar with in the first section of this book. The sight reduction tables created in this manner represent a fraction of the infinite number of solutions available by using the above equations. The altitude and azimuth equations are the basis for sight reduction work with a scientific calculator. The applications and methods will be discussed in detail in the next chapter.

With the observed altitude, computed altitude, and azimuth known, the line of position is plotted from the assumed position. In 1875 the French navigator *Marcq St. Hilaire* documented the altitude-intercept method of plotting a line of position. The altitude-intercept method is still the preferred technique used today.

Figure 14.7 shows the arrival of parallel rays from a body to the earth's surface. The distances between the two different observed angles and the computed angle have been exaggerated on the earth's surface to illustrate the altitude-intercept concept.

Two or more lines of position are used to determine the terrestrial position of an observer on the earth's surface. The terrestrial postion of an observer is called a simultaneous fix or a running fix, as previously discussed.

If you have difficulty with any of this material, review the chapter several times. If you still have difficulties, go to the next chapter and immerse yourself in the practical applications contained

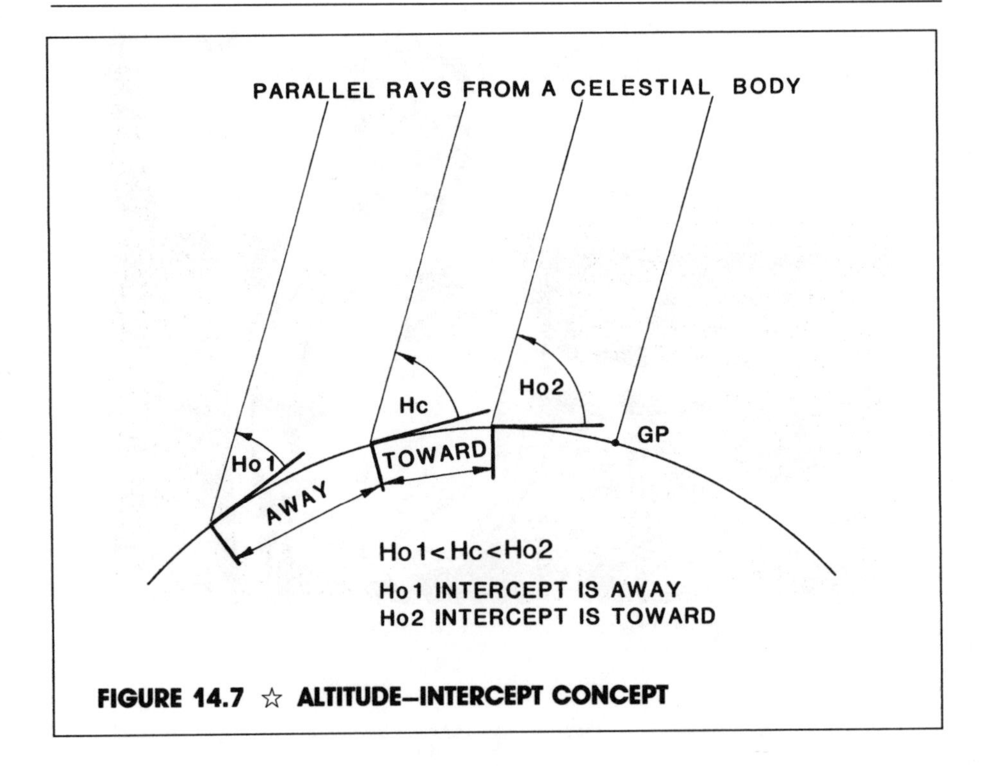

FIGURE 14.7 ☆ ALTITUDE–INTERCEPT CONCEPT

there. Once you have learned solutions of the navigational triangle, you will surely begin to understand the theory. The actual application of this material has been greatly simplified by the use of a scientific calculator, eliminating long and laborious calculations.

DIAMOND HEAD ASTERN
HONOLULU, HI

FIFTEEN

INTERCEPT-AZIMUTH

Prior to the availability of sight reduction tables, navigators laboriously solved complex trigonometric formulas with logarithms. Logarithms are cumbersome, difficult to use, and take a lot of time and effort. However, when solving the navigational triangle with logarithms, the figure values were added and subtracted, as opposed to the more difficult practice of multiplying sines, cosines, and tangents of angles of five or more significant figures.

Today, with inexpensive scientific calculators available, complex sight reduction computations can be performed with ease, provided that principles of celestial navigation are clearly understood.

I've been doing my sight reduction work with a scientific calculator for the past few years. Because of my strong mathematical background, I could have learned calculator navigation at the outset. However, inexpensive scientific calculators were not available in the sixties, so slide rule and tabular solutions were preferred at that point.

The sun sights taken on September 1, 1986, the first day of our voyage from California to Hawaii, will be used for the first example. I could have selected a different problem set, but I felt it important for you to be able to compare the solutions arrived at by the two different methods, and experience the advantages of using a calculator. I also felt that a problem you've already solved would be easier to visualize. Go to Figures 3.3, 4.3, and 4.4, or your completed worksheets.

The first sun sight is taken at 19h Om 12s GMT. Corrections are made to Hs and Ha, and Ho is determined to be 59°00′. The DR longitude and latitude are 122°43′ W and 34°53′ N respectively.

One of the major advantages in calculator navigation is that only one assumed position need be selected for all the sights required to determine a terrestrial position. In some cases I've used the DR position for the AP. In most cases, and for convenience, I prefer to use the longitude and latitude in degrees and zero minutes closest to the DR position.

Begin by finding the meridian angle (t) as follows:

GHA Hours	105°00.7′
Min. & Sec.	0°03.0′
Total GHA	105°03.7′
Ass. Long.	123°
t (E)	17°56.3′

Note that the meridian angle is to the east. Think of the GP as being at 105°03.7′ W and the AP at 123° W. The GP is surely east of the AP of the observer.

The declination is 8°11′ N.

Unlike LHA, do not be concerned with adjusting longitude so that t is in degrees without minutes. Because t can be calculated in degrees and minutes, it is possible to adjust the meridian angle so only one assumed position is selected for all of the sights used in determining a terrestrial position. One assumed position saves a considerable amount of plotting effort and time.

In order to use a scientific calculator it is necessary to convert expressions of degrees and minutes into degrees and decimal fractions. The conversion is achieved by dividing the minutes portion by 60 minutes with your calculator as follows:

$t\ (E) = 17°56.3' = 17° + 56.3/60 = 17.938°$

$d = N\ 8°11' = N\ 8° + 11/60 = N\ 8.183°$ where d is declination

$L = N\ 35°$ where L is assumed latitude

The following equations are utilized to solve the navigational triangle for the computed altitude (Hc) and the azimuth (Z) for all celestial bodies and have been rewritten and bracketed to facilitate computation with a scientific calculator:

Equation 1: $Hc = \sin^{-1}[(\sin L \times \sin d) + (\cos L \times \cos d \times \cos t)]$

and

Equation 2: $Z = \cos^{-1}\{[\sin d - (\sin L \times \sin Hc)]/(\cos L \times \cos Hc)\}$

When declination (d) is contrary to the observer's elevated pole, d is entered as a negative number. On most calculators the negative value of d is entered by first entering the number and then depressing the +/− key.

Latitude (L) is always a positive number because L is always in the same hemisphere as the observer's elevated pole.

Meridian angle (t) is always a positive number.

Begin by inserting the values of L, d, and t into Equation 1 as follows:

$Hc = \sin^{-1}[(\sin 35° \times \sin 8.183°) + (\cos 35° \times \cos 8.183° \times \cos 17.938°)]$

I will attempt to provide information that can be used by the majority of scientific calculators, to reduce the first sight.

In addition to the usual keys for operations like add, subtract, multiply, and divide, there is a key labelled: F or inv. The F or inv key is used to perform the inverse operations of the sin, cos, and tan. The inverse operations are: $\sin^{-1}$, $\cos^{-1}$, and $\tan^{-1}$.

Follow the steps in solving Equations 1 & 2:

Step	Enter	Remarks
1	(	overall bracket
2	(	bracket for 1st operation
3	35	the latitude (L)
4	sin	calculates sin L
5	×	multiply
6	8.183	the declination (d)
7	sin	calculates sin d
8	)	completes first operation
9	+	add
10	(	bracket for 2nd operation
11	35	the L
12	cos	calculates cos L
13	×	multiply
14	8.183	the d
15	cos	calculates cos d
16	×	multiply
17	17.938	the t
18	cos	calculates cos t
19	)	completes 2nd operation
20	)	overall bracket
21	inv or F	for inverse
22	sin	gives Hc.

The display reads: 58.543731 or Hc = 58.544°.

The intercept is calculated by subtracting Hc from Ho. In prior chapters, GOAT, or greater observed angle toward, was used to determine whether the intercept was toward or away. When subtracting with a calculator, Hc is always subtracted from Ho, and if the remainder is positive the intercept is toward, and if the remainder is negative the intercept is away. Ho = 59°00′ = 59.000°. Therefore, the intercept is: Ho − Hc = 59.000° − 58.544° = 0.456° × 60 = 27 nm toward.

Insert the values of d, L, and Hc into Equation 2 and solve for Z as follows:

$$Z = \cos^{-1}\{[\sin 8.183° - (\sin 35° \times \sin 58.544°)]/(\cos 35° \times \cos 58.544°)\}$$

Step	Enter	Remarks
1	(	
2	(	

3	8.183	the d
4	sin	
5	−	subtract
6	(	
7	35	the L
8	sin	
9	×	multiply
10	58.544	the Hc
11	sin	
12	)	
13	)	
14	÷	divide
15	(	
16	35	the L
17	cos	
18	×	multiply
19	58.544	the Hc
20	cos	
21	)	
22	)	
23	inv or F	
24	cos	

The display reads: 144.25599 or Z = 144° rounded to a practical azimuth. It is not possible to plot Z without converting the azimuth to the true azimuth (Zn). There are four possible cases for the determination of Zn, as shown in Figure 15.1.

Case 1: AP is N, t is E and Zn = Z.
Case 2: AP is S, t is E and Zn = 180° − Z.
Case 3: AP is S, t is W and Zn = 180° + Z.
Case 4: AP is N, t is W and Zn = 360° − Z.

The GP can be in either hemisphere and will affect the shape of the navigational triangle. The GP, however, has no effect on the selection of the true azimuth.

The solution for Zn is Case 1 because AP is N, t is E, and hence Zn = Z = 144°.

Plot the resultant LOP as shown in Figure 15.2.

A second sight of the sun is taken at 20h 19m 53s GMT and Ho is determined to be 63°28′. The DR longitude and latitude are 122°50′ W and 34°51 N, respectively.

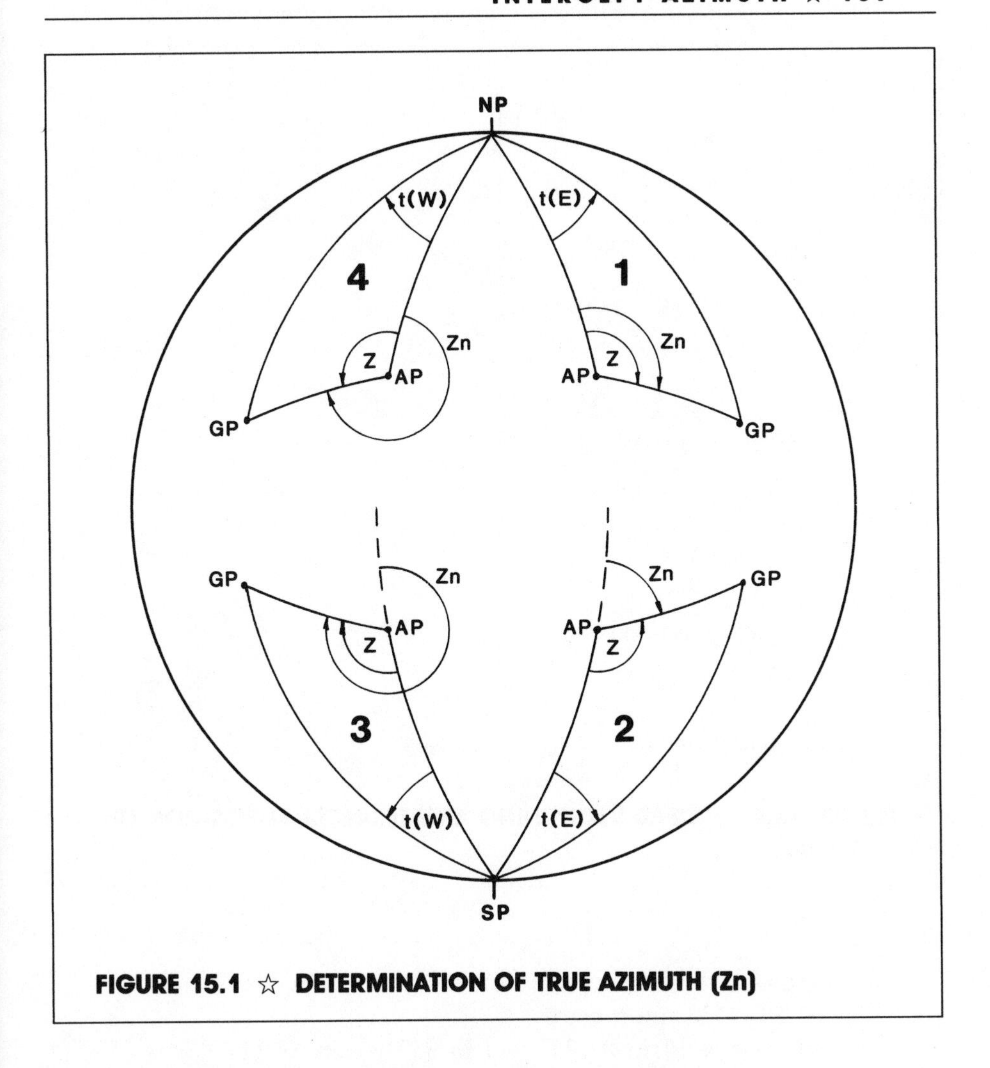

FIGURE 15.1 ☆ DETERMINATION OF TRUE AZIMUTH (Zn)

Reduce the sight and plot the resultant LOP.

GHA Hours	120°00.9′	L = 35° N
Min. & Sec.	4°58.3′	d = 8°10.4′ N = 8.173° N
Total GHA	124°59.2′	Ho = 63°29′ = 63.483°
Ass. Long.	123° W	
t (W)	1°59.2′ = 1.987°	

$Hc = \sin^{-1}[(\sin 35° \times \sin 8.173°) + (\cos 35° \times \cos 8.173° \times \cos 1.987°)]$

$Hc = 63.111°$

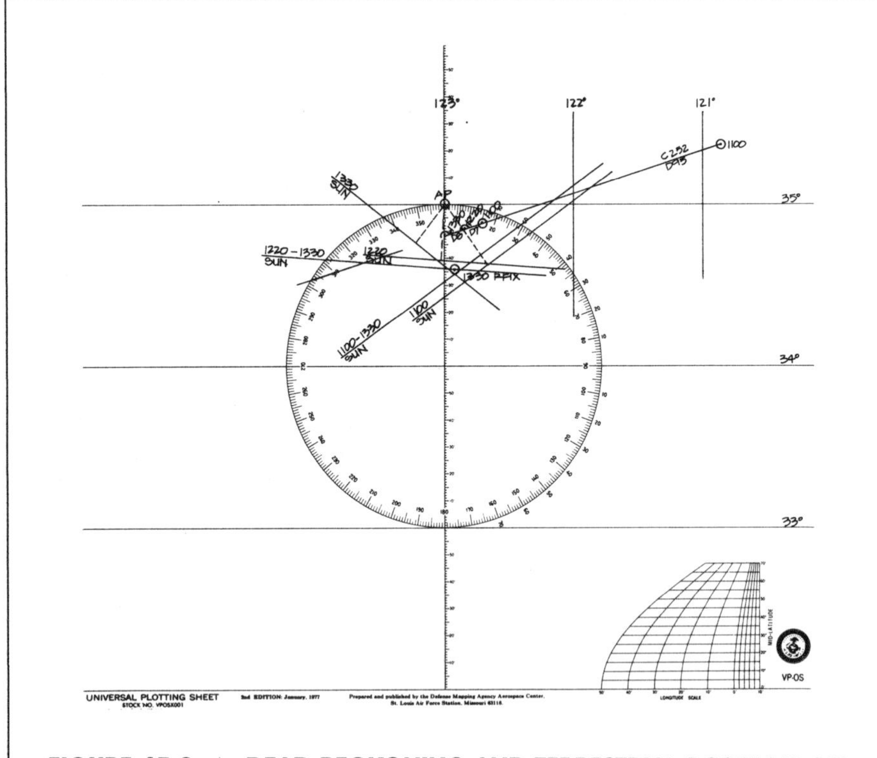

FIGURE 15.2 ☆ DEAD RECKONING AND TERRESTRIAL POSITION ON 9/1/86

Therefore, the intercept is: Ho − Hc = 63.483° − 63.111° = .372° × 60 = 22 nm toward.

$Z = \cos^{-1}\{[\sin 8.173° - (\sin 35° \times \sin 63.111°)]/(\cos 35° \times \cos 63.111°)\}$

$Z = 175.6° = 176°$

Case 4 applies because t is W, AP is N, and therefore, Zn = 360° − Z = 360° − 176° = 184°.

The third sight is taken at 21h 29m 5s GMT and Ho is determined to be 58°10′. The DR longitude and latitude are 123°00′ W and 34°48′ N.

Reduce the sight and plot the resultant LOP.

Check answers with the following.

t (W) = 19° 17.4′, L = N 35°, d = N 8°09.6′, Hc = 57.860°, Ho − Hc = 18′ T, Z = 142° and Zn = 218°

Advance LOP Nos. 1 & 2 to the position they would occupy at the time of LOP No. 3, and compare the terrestrial position obtained by using a scientific calculator to the position obtained by using the sight reduction tables earlier. Note that there are no differences in longitude and latitude between positions obtained by the two methods.

The following examples will help you to learn the techniques for cases of contrary declination and different elevated poles:

1) L = N 30°, d = S 20°, Lo = W 150°, and GHA = 100°. Therefore, t(E) = 50°

or 2) L = S 30°, d = N 20°, Lo = W 150°, and GHA = 100°. Therefore, t(E) = 50°

or 3) L = S 30°, d = N 20°, Lo = W 150°, and GHA = 200°. Therefore, t(W) = 50°

or 4) L = N 30°, d = S 20°, Lo = W 150°, and GHA = 200°. Therefore, t(W) = 50°

$Hc = \sin^{-1}[(\sin 30° \times \sin - 20°) \times (\cos 30° \times \cos - 20° \times \cos 50°)] = 20.615°$

$Z = \cos^{-1}\{[\sin - 20° - (\sin 30° \times \sin - 20°)]/(\cos 31° \times \cos - 20°)\} = 130°$

1) t(E) and AP(N). Therefore, Zn = Z = 130°

2) t(E) and AP(S). Therefore, Zn = 180° − Z = 180° − 130° = 50°

3) t(W) and AP(S). Therefore, Zn = 180° + Z = 180° + 130° = 310°

4) t(W) and AP(S). Therefore, Zn = 360° − Z = 360° − 130° = 230°

If you are experiencing difficulty in visualizing solutions to the examples, I suggest that you draw sketches of the spherical triangles.

If you are having calculation problems, it is probably because declination in the opposite hemisphere is not entered as a negative number. A negative number is entered first by entering the number, then depressing the +/− key, followed by depressing the sin, cos, or tan key.

Using Equations 1 & 2, sights of the sun, moon, stars, and planets can be reduced for plotting. The two equations replace the sight reduction table in the almanac and provide an infinite number of solutions. Sight reduction tables are limited in the number of solutions, and a variety of assumed positions are used to compensate for limitations.

Though the selection is limited, *Vol. 1, Pub. No. 249, Sight Reduction Tables for Air Navigation* is still the fastest method of selecting and observing stars available today. The sight reduction effort required is minimal and I utilize these tables frequently. Electronic devices and batteries have a way of failing at the wrong time, so don't throw away any of your sight reduction tables. The tables should be kept in a dry space on your craft for backup purposes.

FAREWELL TO FRIENDS
MORRO BAY, CA

SIXTEEN

LATITUDE AND LONGITUDE

LATITUDE AT LAN

As mentioned in a previous chapter, an LAN sun sight can be a very accurate method of determining latitude. Let's explore the possibility of using a calculator to determine latitude at LAN.

Begin with Equation 1 as follows:

$$Hc = \sin^{-1}[(\sin L \times \sin d) + (\cos L \times \cos d \times \cos t)]$$

At LAN, the AP of the observer and the GP of the sun are on the same meridian. Therefore, $t = 0°$, $\cos t = \cos 0° = 1$, and Equation 1 becomes:

$Hc = \sin^{-1}[(\sin L \times \sin d) + (\cos L \times \cos d)]$

The bracketed mathematic statement is a trigonometric subtraction formula used to further reduce the equation to:

Equation 3: $Hc = \sin^{-1}[\cos (L - d)]$

An advantage to Equation 3 is the LOP is determined by the intercept-azimuth method, and the LOP is plotted as a horizontal line. Another advantage is that different cases are easily visualized. Some examples are as follows:

L = N 30° and d = N 20°. Therefore, Zn = 180° or
L = S 30° and d = S 20°. Therefore, Zn = 0°
$Hc = \sin^{-1}[\cos(30° - 20°)] = \sin^{-1}[\cos 10°] = 80°$
L = N 20° and d = N 20°. Therefore, no Zn (point of position) or
L = S 20° and d = S 20°. Therefore, no Zn (point of position)
$Hc = \sin^{-1}[\cos(20° - 20°)] = \sin^{-1}[\cos 0°] = 90°$
L = N 10° and d = N 20°. Therefore Zn = 0° or
L = S 10° and d = S 20°. Therefore Zn = 180°
$Hc = \sin^{-1}[\cos(10° - 20°)] = \sin^{-1}[\cos -10°] = 80°$
L = N 30° and d = S 10°. Therefore, Zn = 180° or
L = S 30° and d = N 10°. Therefore, Zn = 0°
$Hc = \sin^{-1}[\cos(30° - -10°)] = \sin^{-1}[\cos 40°] = 50°$.

In the last calculation a negative number was subtracted from a positive number. The rule is that when a negative number is subtracted from a positive number, the numbers are added. The calculator will keep track of subtracting negative numbers for you. The last example is solved as follows: depress: (key; depress: (key; enter: 30°; depress: − key; enter: 10°; depress: +/− key; depress:) key; depress: cos key; depress:) key; depress: inv or F key; and depress: sin key. The answer in the display is 50°.

On September 5, 1986, at approximately local noon, the DR longitude and latitude is 130°16′ W and 32°26′ N, respectively, as shown in Figure 16.1.

Calculate the GHA at LAN and the declination as follows:

GHA at 20h GMT	= 120°20.4′	Dec. 6°41.8′ N	L = 32°26′ N
39m 42s	= 9°55.5′	−0.6′	L = 32.433° N
20h 39m 42s	= 130°15.9′	d = 6°41.2′ N	= 6.687° N

With minutes to spare, the sun is sighted at its greatest altitude. The altitude is corrected and Ho is determined to be 64°13′.

Insert the values of L and d into Formula 3 as follows:

$$Hc = \sin^{-1}[\cos(32.433° - 6.687°)] = \sin^{-1}[\cos 25.746°]$$

$$Hc = 64.254° = 64°15'$$

The intercept is calculated as follows:

$$Ho - Hc = 64°13' - 64°15' = 2 \text{ nm away}$$

Because the GP of the sun is due south of the observer's AP, Zn = 180°, and the intercept of 2 nautical miles is added to the DR Lat. to obtain the Lat. as follows:

$$\text{Lat.} = 32°26' \text{ N} + 2' = 32°28' \text{ N}$$

The most important factor in accurately determining latitude by Equation 3 is to be certain that the sun is sighted at its maximum altitude. The maximum altitude is always when the sun is on the observer's meridian, and the sun is said to be in *upper transit*. When the sun is on the opposing meridian on the opposite side of the

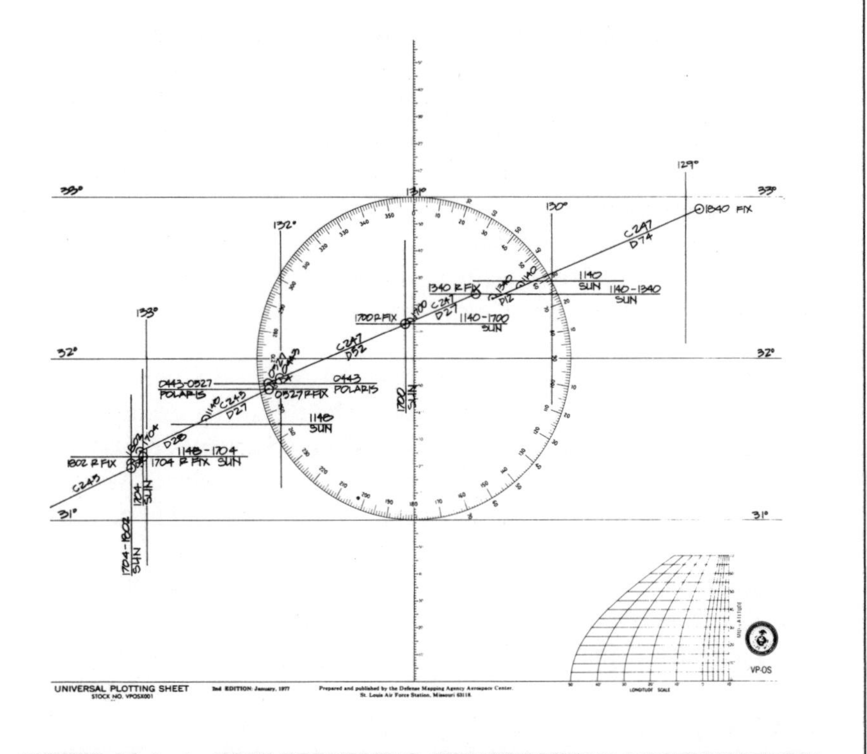

FIGURE 16.1 ☆ DEAD RECKONING AND TERRESTRIAL POSITIONS FROM 9/4/86 TO 9/6/86

earth, at the observer's local midnight, the sun is said to be in *lower transit*.

Longitude, unlike latitude, is completely dependent on accurate knowledge of time. All calculations involving longitude will require the conversion of time to the equivalent angular measurement.

LONGITUDE BY TIME SIGHT

The *time sight* is the traditional method of calculating to determine longitude. Begin with Equation 1 as follows:

$$Hc = \sin^{-1}[(\sin L \times \sin d) + (\cos L \times \cos d \times \cos t)]$$

The terms can be algebraically manipulated for calculating t as follows:

Equation 4: $t = \cos^{-1}\{[\sin Hc - (\sin L \times \sin d)]/(\cos L \times \cos d)\}$

If Hc, L, and d are accurately known, it is possible to calculate t. Knowing t and the GHA of the body, determined at the time of the observation, the longitude (Lo) can be determined by the following formula:

Equation 5: $Lo = GHA \pm t$

Do not use the time sight if latitude is not accurately known. Mistakes in latitude create large errors in t, and hence large errors in longitude.

The latitude is usually calculated at LAN and longitude is determined about two hours later, as in the following example:

Approximately two hours after LAN on September 5, 1986, at 22h 40m 32s GMT, a second sight of the sun is taken. After corrections were applied to the altitude, Ho is determined to be 52°06′. The DR course line is extended 12 miles on a bearing of 247°. The latitude from LAN, a horizontal LOP, is advanced along the DR course to the later time and is determined to be: 32°23′ N.

To utilize Equations 4 & 5, calculate GHA and d as follows:

GHA at 22h GMT	= 150°20.8′	Dec. = 6°40.0′ N	L = 32°23′ N
40m 32s	= 10°08.0′	−0.6′	L = 32.383° N
22h 40m 32s	= 160°28.8′	d = 6°39.4′ N	= 6.657° N
		Ho = 52°06′ = 52.10°	

Insert Ho, L, and d into Equation 4 obtaining:

$t = \cos^{-1}\{[\sin 52.10 - (\sin 32.383 \sin 6.657)]/(\cos 32.383 \cos 6.657)\}$

$t(W) = 29.921° = 29°55'$

Insert GHA and t into Equation 5 obtaining:

$Lo = GHA - t = 160°29' - 29°55' = 130°34'$ W

"t" is subtracted from GHA because the GP of the sun is to the west of the AP of the observer.

The longitude is plotted by drawing a circle around the point on the horizontal LOP advanced to 1440 hours at the determined longitude, as shown in Figure 16.1. The position is labelled: 1340 R FIX and a new DR is drawn bearing 247°.

LONGITUDE BY PRIME VERTICAL SIGHT

A line of position plotted on a true azimuth of 90° or 270° is the longitude, or a line of longitude. Recall that a great circle passing through the observer's zenith, nadir, and a body is called a vertical circle. A vertical circle through the observer's east-west horizon is called the *prime vertical circle.* The problem is to determine when a celestial body will be on the prime vertical. It is important to realize that not all bodies reach the prime vertical. For a body to reach the prime vertical it is necessary for the latitude and declination to be of the same name, and for the declination to be less than the latitude. Applying the law of cosines and the trigonometric reduction formulas, we can write an equation for d as follows:

$\sin d = \sin L \times \sin Hc + \cos L \times \cos Hc \times \cos Z$

When Z is 90° or 270°, due east or west, the cos Z becomes 0, and the last term of the above formula disappears. Solving the equation for Hc we obtain:

Equation 6: $Hc = \sin^{-1}(\sin d/\sin L)$

Using the *law of sines for spherical trigonometry*, a formula for azimuth is derived from the navigational triangle as follows:

$\sin (90° - d)/\sin Z = \sin (90° - Hc)/\sin t$

Substituting reduction formulas for co-angles we obtain:

$\cos d/\sin Z = \cos Hc/\sin t$

Rearranging the terms and solving for sin Z we obtain:

$\sin Z = (\sin t \times \cos d)/\cos Hc$

Caution! Do not use the above equation for calculating an azimuth. Equation 2 may appear to be more complicated. However, when using Equation 2, the direction of Z is not as confusing to interpret as it is if the above equation is used.

If Z is equal to 90°, or 270°, the sin of Z becomes 1 and the formula can be arranged as follows:

Equation 7: $t = \sin^{-1}(\cos Hc/\cos d)$

Equation 7 is used to solve for the meridian angle at the exact time the body is on the prime vertical.

Combining the meridian angle with GHA, Equation 5 is selected to determine the longitude.

On September 5, 1986 a *prime vertical sight* of the sun is taken. The course line from the 1340 position is advanced 27 nautical miles to the projected 1700 position. The DR longitude and latitude are 131°02′ W and 32°13′ N, respectively.

Find the longitude:

At 1700 9/5/86	d = 6°36.3′ N = 6.605° N
0900 from Greenwich	L = 32°13′ N = 32.217°
2600 Greenwich	
− 2400	
0200 GMT 9/6/86 (next day)	

Inserting d and L into Equation 6:

$Hc = \sin^{-1}(\sin 6.605°/\sin 32.217°) = 12.46° = 12°27.6'$

The sextant is preset to Hs. The sun is observed periodically in order to take the sight at the exact time when the sun is observed at the preset altitude. Note, the corrections are applied in reverse. That is, corrections that are normally added are subtracted, and corrections that are normally subtracted are added:

Ho	12°27.6′
Add'l Corr	−0.1′
Alt. Corr.	−11.7′ LL
Ha	12°15.8′
IC	−
Dip	3.4′
Hs	12°19.2′ = 12°19′

With the sextant preset at 12°19′, the sun is observed at 2h 0m 23s GMT.

"t" is calculated by inserting Hc and d into Equation 7 as follows:

$$t(W) = \sin^{-1}(\cos 12.46°/\cos 6.605°) = 79.411° = 79°24.7'$$

The GHA is calculated for 2h 0m 23s GMT on 9/6/86 as follows:

GHA 2h	210°21.6′
0m 23s	0°05.8′
GHA 2h 0m 23s	210°27.4′

Insert GHA and t into Equation 5 to compute the longitude as follows:

$$Lo = 210°27.4' - 79°24.7' = 131°03' \text{ W}$$

To determine the new position the longitude is plotted as a vertical line on the plotting sheet. The intersection of the latitude, advanced to the time of the prime vertical sight, is the terrestrial position and is labelled: 1700 R FIX.

The rate of change of altitude of a celestial body is a significant factor in determining the accuracy of a sight of the body. The rate of change of altitude of a body is at a maximum when on the prime vertical, and that rate of change is significantly more than when a body is near the observer's meridian. A prime vertical observation yields a significantly more accurate determination of longitude than if the body is observed near the observer's meridian.

LONGITUDE BY TIMING SUNRISE AND SUNSET

Longitude can be determined by accurately timing *sunrise* or *sunset*. In practice, the horizon must be clearly visible and unobscured by clouds to use the sunrise-sunset method.

One of the major advantages (or disadvantages) of timing sunrise and sunset is that a sextant is not required. Because a sextant is not required, the method is recommended as a survival technique for lifeboat navigation.

I do not recommend practicing this technique without a sextant if you don't have to. Viewing the sun directly can be extremely hazardous to your eyes. I have observed sunrise and sunset with a sextant adequately shaded and set at 0°. If you decide to observe the sun at sunrise or sunset, please use some method of shading, or a sextant, as I am recommending.

When the sun or any celestial body is first or last seen at the horizon, it is actually 34.5′ below the horizon. The 34.5′ represents refraction and standard conditions. The upper limb of the sun is 34.5′ below the horizon and the center of the sun is an additional distance called the *semidiameter* (SD) below the upper limb at the horizon. The value of SD is found at the bottom of the column headed: SUN in the daily right hand white pages of the *Nautical Almanac*. The average SD is 16′ and close enough for our calculations. Adding 34.5′ to 16′ we obtain the calculated altitude of the sun: 50.5′ below the horizon. Therefore, Hc = −50.5′ = −50.5′/60 = −0.84° and is inserted into Equation 4. The equation is mathematically rearranged and solved for t as follows:

Equation 8: $t = \cos^{-1}\{[\sin -0.84° - (\sin L \times \sin d)]/(\cos L \times \cos d)\}$

GHA of the sun is calculated for the exact time the upper limb of the sun is observed on the horizon. Longitude is determined by inserting GHA and t into Equation 5.

On September 6, 1986, at 0443, Polaris is observed at twilight and the latitude is determined to be 31°51′ N. The upper limb of the sun is observed on the horizon at 14h 26m 45s GMT. The DR course line is extended to 0527, a distance of 4 nautical miles on a bearing of 247°, as shown in Figure 16.1. The DR longitude and latitude are 132°06 W and 31°51′ N, respectively. The LOP for Polaris is advanced from 0443 to 0527, a distance of 4 nautical miles in the direction of the DR course, and the latitude is determined to be 31°48′ N.

The GHA and d are obtained as follows:

9/6/86 14h	30°24.1′	Dec. 6°25.1′ N
26m 45s	6°41.3′	−0.4′
GHA 14h 26m 45s	37°05′	d = 6°24.7 N
		d = 6.412° N

L = 31°48′ N = 31.8′ N

We insert L and d into Equation 8 as follows:

$t(E) = \cos^{-1}\{[\sin - 0.84 - (\sin 31.8 \times \sin 6.412)]/(\cos 31.8 \times \cos 6.412)\}$

$t(E) = 94.985° = 95°00'$

We insert t and GHA into Equation 5 as follows:

$Lo = 37°05' + 95°00' = 132°05'$

The sunrise-sunset technique of obtaining longitude can be used only if the latitude is accurate. This technique is recommended for use in lower latitudes, and tends to be considerably less accurate with increasing latitude. The longitude is plotted as a point on the LOP of Polaris previously advanced to the time of the sight.

LATITUDE BY SUNSET SIGHT

It is possible to find the latitude by taking a sight of the sun at actual sunset if the longitude is known. If the observation is made when the center of the sun is on the horizon, the altitude is 0°. When the altitude is 0°, sin Hc = sin 0° = 0 and Formula 1 becomes:

$0 = \sin L \sin d + \cos L \cos d \cos t$

and

$\cos t = -\sin L \sin d/\cos L \cos d = -\tan L \tan d$

and

$L = \tan^{-1}[-(\cos t/\tan d)]$ Formula 9

The altitude for observing the sun can be extracted from the almanac by going to the first page following the inside front cover headed: ALTITUDE CORRECTION TABLES 0°–10°—SUN, STARS, PLANETS A3.

Go down the "Apr–Sept" (right hand) App. Alt. column and the Lower Limb column simultaneously to App. Alt. − 0°15′. The corresponding lower limb correction is − 15.3′. Adding the two values, the sum becomes approximately 0°. The 0° indicates that the center of the sun is on the horizon if the observed altitude is about 0°15′.

On September 6, 1986, at approximately 1704, the sun is observed on the prime vertical. The earlier observation at LAN is advanced to 1704, and the terrestrial position is determined to be 31°24′ N and 133°02′ W. The DR course line is drawn at 245° and extended 5 nautical miles to the time of sunset at approximately 1802. The DR Lat. and Long. are 31°22′ N and 133°07′ W, respectively. The sextant is preset to 18′(altitude plus dip) and the sun is observed as the lower limb touches the horizon at 3h 5m 52s GMT 9/7/86. The GHA and d are calculated as follows:

9/7/86 3h	225°26.9′	Dec 6°13.0′ N
5m 52s	1°28.0′	−0.1′
3h 5m 52s	226°54.9′	d = 6°13.0′ N = 6.217° N

L = 31°22′ = 31.367° N

We insert GHA and Lo into Equation 5 and solve for t as follows:

$$t = GHA - Lo = 226°55' - 133°07' = 93°48' = 93.8°$$

We insert t and d into Formula 9 and solve for L as follows:

$$L = \tan^{-1}[-(\cos 93.8°/\tan 6.217°)] = N\ 31.316° = N\ 31°19'$$

The terrestrial position is determined to be 31°19′ N and 133°07 W. Determining latitude at sunset is not as well known as the traditional method of determining longitude at sunset. The latitude at sunset technique should not be used unless the longitude is well known. A small error in longitude will create a much larger error in the calculation of latitude. Because of the larger magnitude of error in the calculated answer, I don't recommend the sunset-latitude sight be used frequently. Just as longitude is more accurately determined at the prime vertical, sunset, or sunrise than at LAN, latitude is more accurately determined at LAN than at sunset.

As discussed earlier, intercept-azimuth methods of sight reduction produce the most reliable determination of terrestrial positions.

The noon sight is extremely accurate for determining latitude. The sight however, cannot be taken if a cloud obscures the sun at LAN.

The time sight, prime vertical sight, and sunrise or sunset sights must be taken at the appropriate times. They are sensitive to weather conditions, and are not always obtainable.

SEVENTEEN

MISCELLANEOUS CALCULATIONS

To navigate with greater precision, and not be limited to observations of the sun around noon, the prudent navigator will want to obtain positions derived from sights of the moon, stars, and planets. Therefore, the subjects of sunrise, sunset, moonrise, moonset, civil twilight, and nautical twilight will be discussed.

The *Nautical Almanac* is used to determine the middle of the working period between civil and nautical twilight. The reason for taking observations during the middle of the working period is that the *navigational planets* and stars of all magnitudes and the horizon are usually visible, provided that the sky is clear. I have observed the brightest (first magnitude) stars before the end of the civil twilight period and also, when visibility was exceptional, and the horizon could be seen, toward the end of nautical twilight.

Twilight or nautical twilight, for celestial navigational purposes, is from sunset, when the sun is last seen at the horizon; until the end of nautical twilight, when the sun is 12° below the horizon; or from the begining of nautical twilight until the end of nautical twilight at sunrise, when the sun first appears at the horizon.

Civil twilight is from sunset to the end of civil twilight when the sun is 6° below the horizon, or from the begining of civil twilight to sunrise.

The easiest method of determining the time for twilight observations is to use the tabulated information on the right side of the right hand white pages of the *Nautical Almanac*. If desired, the time for twilight observations can be calculated more accurately by using certain equations.

The following equation is used to determine the local mean time of sunrise, sunset, civil twilight, and nautical twilight in hours as follows:

Equation 10: $\text{LMT(mean sun)} = (180° \pm t)/15$

The sign is plus (+) for sunset, and sunset civil or nautical twilight. The sign is minus (−) for sunrise and sunrise civil or nautical twilight.

Angle t in Equation 10 for sunrise and sunset is derived from Equation 8:

$t = \cos^{-1}\{[\sin -.84 - (\sin L \times \sin d)]/(\cos L \times \cos d)\}$

The equation for t of civil twilight is created by inserting −6° into Equation 4 is as follows:

Equation 11: $t = \cos^{-1}\{[\sin - 6° - (\sin L \times \sin d)]/(\cos L \times \cos d)\}$

The equation for t of nautical twilight is created by inserting −12° into Equation 4 as follows:

Equation 12: $t = \cos^{-1}\{[\sin 12° - (\sin L \times \sin d)]/(\cos L \times \cos d)\}$

After t has been determined, applied to Equation 10, and *local mean time* (LMT) is determined, one additional correction is applied to obtain the local time at the full-hour meridian. The additional correction is extracted from the bottom right hand side of the daily white pages in the almanac, and is called the *equation of time*.

The equation of time is used to compensate for the apparent lead or lag in the time of the sun at *meridian transit*. What this means is that the earth does not rotate in a perfect 24-hour circle, and the time when a meridian passes the center of the sun has to be changed to compensate for the lead or lag of the meridian with respect to the sun.

If the true sun is ahead of schedule, the equation of time is subtracted from the LMT (*mean sun*). If the true sun is behind schedule, the equation of time is added to LMT (mean sun). The result is the LMT(true sun). It would be wonderful if the sun and the other celestial bodies would behave perfectly in step with mean time. If celestial bodies would behave in this manner, a tremendous amount of information could be eliminated from the almanac.

The equation of time was not required to solve for longitude because the GHA of the sun, as printed in the *Nautical Almanac*, takes the lead or lag of the sun at the mean time of the sight into consideration.

When calculating sunrise or sunset we are using mathematical equations and not actual data from the almanac. Hence, the equation of time, extracted from the *Nautical Almanac*, is needed to compensate for the sun's actual path.

On September 4, 1986, in the late afternoon, we decided to compile a list of stars for possible observation at twilight. We also wanted to determine our longitude at sunset and be able to observe some of the brighter stars earlier in the twilight period. Our DR position was advanced to approximate twilight, and was approximately 128°56′ W and 32°55′ N = 32.917° N.

Calculate sunset, civil twilight, and nautical twilight.

From the *Nautical Almanac*, the declination is approximately 6°56′ N = 6.933° N between 0300 and 0400 GMT on 9/5/86, the next day at Greenwich.

We insert L and d into Equation 8 as follows:

$t = \cos^{-1}\{[\sin - .84 - (\sin 32.917 \times \sin 6.933)]/(\cos 32.917 \times \cos 6.933)\}$

$t = 95.5267°$

We insert t into Equation 10 as shown:

LMT(mean sun) = $(180° + 95.5267°)/15 = 18.3684$ hours

A fast method of converting hours to hours, minutes, and seconds is by using the calculator as follows:

18.3684 h, $18.3684 - 18 = .3684$, $.3684 \times 60 = 22.104$m, $22.104 - 22 = .104$, $.104 \times 60 = 6.2 = 6$s. Therefore, LMT(mean sun) = 18h 22m 6s

Go to the right hand bottom portion of the daily white pages in the almanac headed: 1986 SEPTEMBER 4, 5, 6 (THURS., FRI., SAT.), or Figure 8.2.

Go to the little box on the bottom of the page headed: SUN.

Go to the column headed: Mer. Pass. Note that the true sun is passing the meridians earlier than noon, at 11h 59m. Because the true sun is ahead of the mean sun, the value of the equation of time is subtracted from the mean time of meridian passage.

Go to the columns headed: Eqn. of Time, and to the column headed: 12h. On the 5th day the value is: 1m 14s and is subtracted from the LMT(mean sun) to obtain the LMT(true sun) as follows:

LMT(true sun) = 18h 22m 6s − 1m 14s = 18h 20m 52s

Go to the column headed: sunset in the almanac. Note that sunset is 18h 22m at N35° and 18h 18m at N30°. The time of sunset calculated by our formula is extremely accurate.

Proceeding with the calculation for civil twilight, L and d are inserted into Equation 11 as follows:

$$t = \cos^{-1}\{[\sin -6° - (\sin 32.917 \times \sin 6.933)]/(\cos 32.917 \times \cos 6.933)\}$$

$$t = 101.7799°$$

We insert t into Equation 10 as follows:

LMT(mean sun) = (180° + 101.7799°)/15 = 18.7853 h = 18h 47m 7s

LMT(true sun) = 18h 47m 7s − lm 14s = 18h 45m 53s

Calculating for nautical twilight, L and d are inserted into Equation 12 as follows:

$$t = \cos^{-1}\{[\sin 12° - (\sin 32.917 \times \sin 6.933)]/(\cos 32.917 \times \cos 6.933)\}$$

$$t = 109.1604°$$

We insert t into Equations 10 as follows:

LMT(mean sun) = (180° + 109.1604°)/15 = 19.2774 h = 19h 16m 39s

LMT(true sun) = 19h 16m 39s − 1m 14s = 19h 15m 25s

Sunrise, civil twilight, and nautical twilight are easily calculated in the same manner.

The formula for t for moonrise and moonset is as follows:

Equation 13: $t = \cos^{-1}\{[\sin(-.57 - SD + HP) - (\sin L \times \sin d)]/(\cos L \times \cos d)\}$

SD is the semidiameter, and HP is the *horizontal parallax*. SD is taken from the three choices at the bottom of the moon column in the daily white pages of the Nautical Almanac. The three choices of SD are for the three calendar days represented on the page. HP is extracted from the almanac at the approximate GMT of moonrise or moonset. SD and HP are converted from minutes to degrees for insertion into the equation. Don't forget to use the equation of time for the moon.

Unless longitude is to be determined at sunrise or sunset I wouldn't calculate sunrise or sunset by formula. The times listed in the almanac are close enough for determining working periods for observations of celestial bodies. Because of the slight chance that you might end up in the liferaft with an accurate wristwatch (set on GMT), a scientific calculator (that works), and a soggy almanac, you would be able to determine your latitude at LAN and your longitude at sunrise or sunset. If you could read the Polaris tables in the almanac or had the white insert card from *Vol 1, Pub. No. 249 Sight Reduction Tables for Air Navigation*, you could determine your latitude from Polaris during the twilight periods prior to sunrise or after sunset. Lucky guy, there you are experiencing one of the worst possible situations but you are still able to navigate.

DIP AND REFRACTION

Corrections for dip and refraction are found in the almanac, as discussed in earlier chapters. However, if you are without the almanac for some reason, equations can be used to determine these corrections.

The equation used to determine dip, in minutes of degrees, is as follows:

Equation 14: $\text{Dip} = -.97 \times \sqrt{\text{Ht of Eye in Feet}}$

Suppose the observer's eye is 10 ft. above the horizon. The dip is calculated by taking the square root of 10 ft. and multiplying by .97 as follows:

$$\text{Dip} = -.97 \times \sqrt{10 \text{ ft.}} = -.97 \times 3.16' = -3.06' = -3.1'$$

For those desiring to use the metric system, one foot is equal to .3048 meters.

There is a useful relationship for determining the visibility of an object. The technique is to use the height of the eye of an observer and the height of an object sighted to determine the dip of the observer and the object. Twenty percent is added to the dip of the observer and the object. The values are added to determine an approximation of the distance that the object would be visible.

Example: A lighthouse is 450 feet above the water atop a cliff. An observer's eye is 9 feet above the water when the top of the lighthouse is sighted on the horizon. What distance is the observer from the lighthouse? Insert the heights into Equation 14 as follows:

$\text{Dip} \times 1.2 = (-.97 \times \sqrt{450}) \times 1.2 = -.97 \times 21.2' \times 1.2 = -24.7$ nm.

$\text{Dip} \times 1.2 = (-.97 \times \sqrt{9}) \times 1.2 = -.97 \times 3' \times 1.2 = -3.5$ nm

The distance from the observer to the lighthouse is the sum of the dip of the observer and the dip of the lighthouse, plus 20% = $24.7' + 3.5' = 28.2$ nautical miles.

The above example illustrates how celestial observations can be made facing the land. The distance to the shore is known to be greater than 3.5 nautical miles (the dip). Therefore, the sea horizon toward the land is suitable for celestial observations.

The equation for the refraction correction in minutes of degrees, for all the celestial bodies, is as follows:

Equation 15: $R = -.97 \times \tan\{Ha - \tan^{-1}[12 \times (Ha + 3)]\}$

Suppose a sight is corrected for index error and dip. The result is the apparent altitude (Ha). A sight is taken of a star and the Ha is $27°36' = 27.6°$.

What is the refraction correction to the altitude for standard conditions? We insert Ha into Equation 15 as follows:

$R = 0.97 \times \tan\{27.6° - \tan^{-1}[12 \times (27.6 + 3)]\} = -1.84'$

You can check this by going to the inside cover of the *Nautical Almanac* and down the App. Alt. column for the stars and planets. The corrections on either side of 27°36′ are −1.9′ and −1.8′. Therefore, −1.84′ is the refraction correction.

The refractive corrections for the sun are calculated for the upper limb by subtracting the semidiameter (SD) from R and lower limb by adding SD to R. The equations are $UL = R - SD$, and $LL = R + SD$.

Now that we've covered a considerable amount of the material in the *Nautical Almanac* and can calculate much of the tabulated data, would we want to dispense with the almanac? My answer is, emphatically, no! The *Nautical Almanac* has simplified all of the above information and tabulated it in easily retrievable order. We still need all the time-related information like GHA of the sun, moon, and planets. The GHA of Aries, SHA, and declination of the stars is needed. The many corrections for the moon are needed and the list goes on ad infinitum. The *Nautical Almanac* is the cruising navigator's most valuable publication.

HUMPBACK WHALE OFF LANAI

EIGHTEEN

THE SAILINGS II

The great circle and mercator sailings are examples of applications well suited to the use of the scientific calculator.

GREAT CIRCLE SAILING

Equations 1 & 2 can be rewritten and reworked for distance and initial course for *great circle sailing* for all latitudes and longitudes as follows:

Equation 16: $D = \cos^{-1}[(\sin L_1 \times \sin L_2) + (\cos L_1 \times \text{Cos } L_2 \times \cos DLo)]$

Equation 17: $C_1 = \{[\sin L_2 - (\sin L_1 \times \cos D)]/(\cos L_1 \times \sin D)\}$

D is the distance in degrees, C_1 is the initial course, L_1 is the departure latitude, L_2 is the destination latitude, and DLo is the difference between the departure and destination longitudes.

If the destination latitude is in a different hemisphere than the departure latitude, the destination latitude is entered as a negative number (analogous to declination).

The actual distance from departure to destination in nautical miles (minutes of arc) is found by multiplying D by 60.

A great circle course is actually sailed by sailing a series of chords (rhumb lines) to the great circle. The easiest way to sail a great circle course, if there are no obstructions en route, is to calculate a new course every time the terrestrial position is determined, for most of the way to the destination.

A rhumb line course is usually sailed the remaining 300 to 500 miles distance to the final destination, depending on your confidence in the accuracy of your navigational calculations.

Suppose you wanted to calculate a great circle course over an extended distance and there were obstructions en route. Prior to setting sail you would calculate waypoints in the course, and be able to see where the great circle course takes you.

The following equations are used for finding the maximum latitude sailed en route, the distance to that maximum latitude, the difference in longitude between the departure longitude, and the longitude of the point of maximum latitude as follows:

Equation 18: $Lv = \cos^{-1}(\cos L_1 \times \sin C_1$

Equation 19: $Dv_1 = \sin^{-1}[1/(\tan Lv \times \tan C_1)]$

Equation 20: $DLov = \cos^{-1}(\cos Dv_1 \times \sin C_1)$

Equation 21: $L_{1i} = \tan^{-1}(\cos Dv_1 \times \tan Lv)$

Where Lv is the latitude of the *vertex* (maximum latitude), Dv_1 is the distance to the vertex from the departure point, DLov is the difference in longitude from the departure to the vertex, and L_{1i} is the latitude of i waypoints.

When sailing east-west, or roughly so, courses, the vertex is somewhere on the course line. When sailing north-south courses, the vertex is somewhere on an imaginary extension of the course line, as shown in Figure 18.1. An exact north-south course is along a meridian, of course, with 0 or 180 degrees constant course. Note that the calculation of vertex is useful only for further calculations.

The imaginary course line extension does not present any special

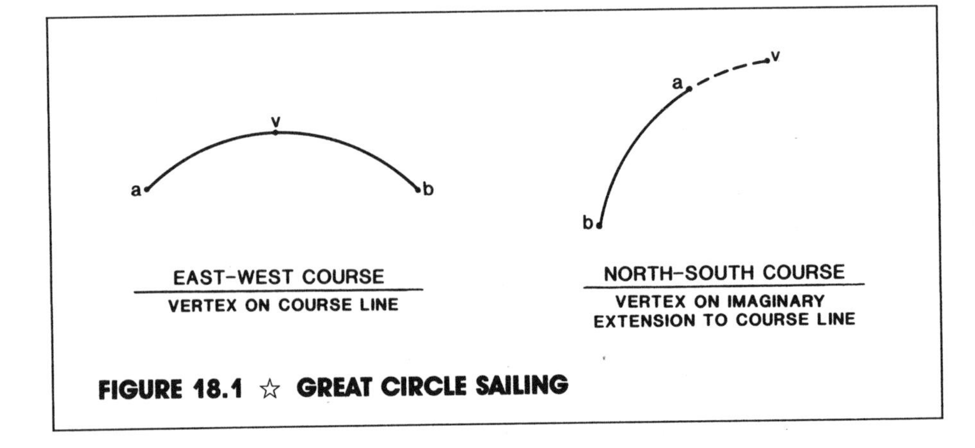

FIGURE 18.1 ☆ GREAT CIRCLE SAILING

problems in waypoint calculations, and the vertex in the example of the great circle course plotted from Morro Bay, California to Hilo, Hawaii is on an imaginary extension of the course line.

Calculate the great circle distance, initial course, and the latitude and longitude of all the waypoints from 125° W to 150° W, in 5° increments as follows:

Morro Bay	Hilo
$L_1 = 35°22'N = 35.367°N$	$L_2 = 19°44'N = 19.733°N$
$Lo_1 = 120°52'W = 120.867°W$	$Lo_2 = 155°04'W = 155.067°W$

$$Dlo = 155.067° - 120.867° = 34.200° = 34°12'$$

Inserting L_1, L_2 and Dlo into Equation 16 we obtain:

$D = \cos^{-1}[(\sin 35.367 \times \sin 19.733) + (\cos 35.367 \times \cos 19.733 \times \cos 34.2)]$

$D = 33.873°$. Therefore, Distance $= 33.873 \times 60 = 2032$ nm

Inserting L_1, L_2, and D into Equation 17 we obtain:

$C_1 = \cos^{-1}\{[\sin 19.733 - (\sin 35.367 \times \cos 33.873)]/(\cos 35.367 \times \sin 33.873)\}$

$C_1 = 108.330°$. Therefore, the true heading: $C_{1n} = 360° - 108° = 252°$.

Inserting L_1 and C_1 into Equation 18 we obtain:

$Lv = \cos^{-1}(\cos 35.367° \times \sin 108.330°) = 39.278°N = 39°17'N$

Inserting Lv and C_1 into Equation 19 we obtain:

$Dv_1 = \sin^{-1}[1/(\tan 39.278° \times \tan 108.330°)] = -23.897° = -23°54'$

Distance to the vertex $= -23.897 \times 60 = -1434$ nm in the opposite direction from the course, and on an imaginary extension of the course line.

Inserting Dv_1 and C_1 into Equation 20 we obtain:

$Dlov = \cos^{-1}(\cos -23.897° \times \sin 108.330°) = 29.786° = 29°47'$

$Lov = Lo_1 + Dlov = 120°52' - 29°47' = 91°05'W$

$V = (39°17'N, 91°05'W)$

The vertex is somewhere in the middle of the United States. The first waypoint en route is calculated at longitude 125° W by adding $DLoa_1 = 125°00' - 91°05' = 33°55' = 33.917°$ to the vertex longitude. The latitude at the first waypoint is calculated by inserting $DLoa_1$ and Lv into Equation 21 and obtaining:

$V + 33°55'$

$La_1 = \tan^{-1}(\cos 33.917° \times \tan 39.278°) = 34.164°N = 34°10'N$

$Loa_1 = 91°05' + 33°55' = 125°00'W$

Waypoint 1 (34°10'N, 125°00'W)

$V + 38°55'$

$La_2 = \tan^{-1}(\cos 38.917° \times \tan 39.278°) = 32.470°N = 32°28'N$

$Loa_2 = 91°05' + 38°55' = 130°00'W$

Waypoint 2 (32°28'N, 130°00'W)

$V + 43°55'$

$La_3 = \tan^{-1}(\cos 43.917° \times \tan 39.278°) = 30.504°N = 30°30'N$

$Loa_3 = 91°05' + 43°55' = 135°00'W$

Waypoint 3 (30°30'N, 135°00'W)

$V + 48°55'$

$La_4 = \tan^{-1}(\cos 48.917° \times \tan 39.278°) = 28.256°N = 28°15'N$

$Loa_4 = 91°05' + 48°55' = 140°00'W$

Waypoint 4 (28°15'N, 140°00'W)

$V + 53°55'$

$La_5 = \tan^{-1}(\cos 53.917° \times \tan 39.278°) = 25.719°N = 25°43'N$

$Loa_5 = 91°05' + 53°55' = 145°00'W$

Waypoint 5 (25°43'N, 145°00'W)

$V + 58°55'$

$La_6 = \tan^{-1}(\cos 58.917° \times \tan 39.278°) = 22.891°N = 22°53'N$

$Loa_6 = 91°05' + 58°55' = 150°00'W$

Waypoint 6 (22°53'N, 150°00'W)

From waypoint 6, draw a rhumb line to the destination.

An example of an east-west course, a great circle course from Yokohama, Japan to San Francisco, California, is calculated as follows:

Yokohama	San Francisco
$L_1 = 35°27'N = 35.45°N$	$L_2 = 37°49'N = 37.817°N$
$Lo_1 = 139°35'E$	$Lo_2 = 122°25'W$

$Dlo = 360° - 139°35' - 122°25' = 98°00'$

$D = \cos^{-1}[(\sin 35.45 \times \sin 37.817) + (\cos 35.45 \times \cos 37.817 \times \cos 98)]$

$D = 74.57°$ Distance $= 74.57 \times 60 = 4474$ nm

$C_1 = \cos^{-1}\{[\sin 37.817 - (\sin 35.45 \times \cos 74.57)]/(\cos 35.45 \times \sin 74.57)\}$

$C_1 = 54.246°$ Therefore, $C_{1n} = 54°$

$Lv = \cos^{-1}(\cos 35.45 \times \sin 54.246) = 48.616°N = 48°37'N$

$Dv_1 = \sin^{-1}[1/(\tan 48.616 \times \tan 54.246)] = 39.376°$

Distance to the vertex $= 39.376 \times 60 = 2363$ nm.

$Dlov = \cos^{-1}(\cos 39.376° \times \sin 54.246°) = 51.148° = 51°09'$

$Lov = 360° - 139°35' - 51°09' = 169°16'W$

V (48°37'N, 169°16'W)

$La_1 = \tan^{-1}(\cos 15° \times \tan 48.616°) = 47.629°N = 47°38'N$

V + 15° (47°38'N, 154°16'W)

$La_2 = \tan^{-1}(\cos 30° \times \tan 48.616°) = 44.505°N = 44°30'N$

V + 30° (44°30'N, 130°16'W)

$La_3 = \tan^{-1}(\cos 45° \times \tan 48.616°) = 38.747°N = 38°45'N$

V + 45° (38°45'N, 124°16'W)

V − 15° (47°38'N, 175°44'E)

V − 30° (44°30'N, 160°44'E)

V − 45° (38°45'N, 145°44'E)

For computing purposes, there is an advantage to the vertex being near the center of the great circle course. The advantage is that the latitude is calculated for two points on either side of the vertex at the same time.

MERCATOR SAILING

In addition to the great circle course, it is customary to plot the mercator, or rhumb line course, on the chart, or charts, of the area sailed.

The rhumb line on a Mercator chart is a straight line between two points and the angle between the course line and all meridians is the same over its entire length. This true course is modified by variation and deviation to get the course to be steered. The rhumb line is not the shortest distance on the earth's surface. The shortest distance is a great circle. As previously discussed, a great circle course curves north of the rhumb line in the northern hemisphere, south of the rhumb line in the southern hemisphere, and intersects with the rhumb line at the equator when crossing from one hemisphere to the next.

For shorter distances and a variety of other reasons, such as obstructions en route, colder weather, and better wind conditions, a Mercator sailing may be the preferred course. See Figure 18.2. A Mercator course can be determined by calculation as well as plotting.

Let's begin by using the rhumb line from Morro Bay to Hilo as an example, as follows:

	Latitude	Meridional Parts
Morro Bay, Ca.	35°22′N	2257.7
Hilo, Hi.	19°44′N	1200.3
	l = 15°38′	1057.4 = m
	l = 15.633 × 60 = 938′	

For Mercator sailing, the course would be the same over the entire length of the plotted course line, and is expressed as follows:

Equation 22: $C = \tan^{-1} (DLo/m)$

Where m is the difference in *meridional parts* from the departure latitude to the destination latitude, and is extracted from *Bowditch, Vol II, Table 5*. Meridional parts are used to compensate for the distortion with the change in latitude on a Mercator chart. DLo is the difference in longitude from the departure point to the destination point in minutes, and C is the course in degrees.

The distance sailed from the departure point to the destination point is expressed for Mercator sailing as follows:

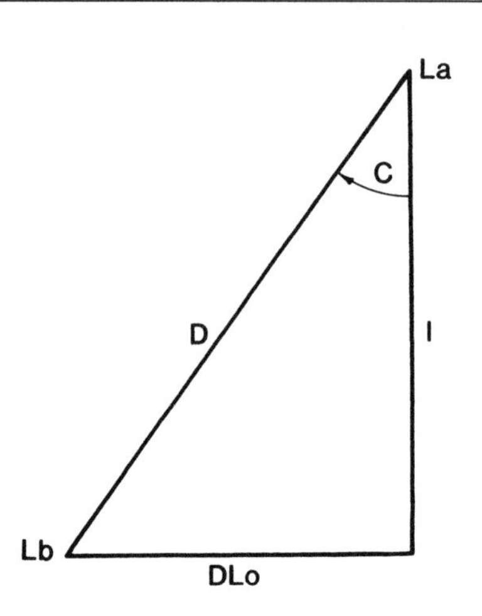

FIGURE 18.2 ☆ MERCATOR SAILING

Equation 23: $D = l/\cos C$

Where l is the difference between the initial and final latitudes in minutes, C is the course in degrees from Equation 22, and D is the distance in minutes, or nautical miles.

$Dlo = 34°12' = 34 \times 60' + 12' = 2052'$

Insert DLo and m into Equation 22 obtaining:

$C = \tan^{-1}(2052/1057.4) =$ S62.738°W. Therefore, $Cn = 180° + 63° = 243°$

Insert l and C into Equation 23 obtaining:

$D = 938/\cos 62.738° = 2048$ nm

The difference between the great circle and Mercator courses was only $2048' - 2032' = 16$ nm. Why sail the great circle course? Recall the description of possible weather conditions in September. The great circle course was approximately 90 nm farther north, and was preferred because of the possibility of encountering tropical cyclonic activity of greater magnitude.

Let's calculate the Mercator course and distance from Yokohama, Japan to San Francisco, California, and compare our findings with the great circle course computed earlier, as follows:

	Latitude	Meridional Parts
San Francisco	37°45′N	2440.0
Yokohama	35°27′N	2263.8
	$1 = 2°18' = 138'$	176.2

$\text{DLo} = 98° \times 60' = 5880'$

$C = \tan^{-1}(5880'/176.2) = 88.284° =$ N 88° E

The difference in distance between the great circle and Mercator course was $4608 - 4474 = 134$ nm.

Since there is only slightly more than two degrees difference in latitude between Yokohama and San Francisco, the approximate latitude of the rhumb line under the vertex of the great circle can be calculated with fair accuracy by devising a simple ratio and proportion equation as follows:

Diff. in latitude $= 1 \times (\text{DLov}/\text{DLo}) = 138 \times (51.148°/98°) = 72' = 1°12'$

Therefore, the approximate latitude of the rhumb line under the vertex of the great circle is $35°27' + 1°12' = 36°39'$ N and the most northerly point reached is thus $11°58' = 11 \times 60 + 58 = 718$ nm farther north by sailing the great circle course, as compared to the Mercator route.

I would rather be at sea an extra day and a half than sail as far north as the Aleutian Islands. The rhumb line could be sailed below 40° N, and warmer temperatures would probably be experienced.

Today, there are a variety of preprogrammed and programmable calculators available. These calculators can further diminish the sight reduction process. Many of these calculators have reduced the effort to inserting data. Some of the calculators eliminate the use of most of the *Nautical Almanac*, depending on the programs available. No knowledge of spherical triangles or precomputed tables is required.

There are wonderful electronic navigational systems such as Loran C and Sat Nav available at reasonable prices to today's cruising sailors. These would completely eliminate the need for celestial navigation if they could be relied upon solely. However, an electronic device is only one of many tools available to you, and a malfunction of the device should be of no special concern to you if you are an adequately prepared navigator. Because in the real world, the world of the small-craft celestial navigator, the world of wet decks, wet cabins, wet everything, extreme motion, dead batteries—and the list goes on ad infinitum—salt water and electronics don't mix.

If you do have a Sat Nav or a Loran C, I recommend that in addition to using the device, you discipline yourself to practice traditional celestial navigation at least every third day at sea, and for the last four or five days consecutively, prior to making a landfall. The routine I am suggesting is your finest backup and will generally keep you out of trouble. And because basic scientific calculators are inexpensive, I recommend you carry several aboard, complete with a replacement supply of batteries.

I also recommend celestial navigation for coastal sailors. I am constantly amazed at how little celestial navigational knowledge coastal sailors have. Fog and adverse weather conditions have a greater effect on coastal cruisers than blue-water sailors. When the weather has deteriorated to minimal visibility, or wind conditions have increased to moderate gale proportions, land should not be approached. The prudent sailor should heave-to on the seaward tack and wait for the weather to change. Once visibility and fair sea conditions have returned, celestial navigation can be used to establish a terrestrial position and reach a safe harbor.

I sincerely hope the information contained in this course will be used to expand your cruising horizons. A successful voyage in a small craft is one of the last great adventures on this planet. I hope you will enjoy cruising and gain the confidence that is acquired only by knowledge and experience.

APPENDIX A

ABBREVIATIONS

A	away (altitude difference)
a_0	Polaris correction
a_1	Polaris correction
a_2	Polaris correction
Add'l	additional
Alt.	altitude
AP	assumed position
App.	apparent
Ass.	assumed
C	course angle (0° to 180° east-west)
C_1	initial course angle (0° to 180° east-west)
C_{1n}	initial true course angle (0° to 360° true)
Cn	true course angle (0° to 360° true)
Corr	correction
cos	cosine
$\cos^{-1}$	inverse cosine
d	declination, interpolation value
D	distance, angular distance
Dec.	declination
Dlo	difference in longitude
DLov	difference in longitude between departure and vertex

DR dead reckoning, dead reckoning position
Dv_1 angular distance to vertex from departure point
E east
ETA estimated time of arrival
ETD estimated time of departure
F Fahrenheit, simultaneous fix
ft. foot, feet
GHA Greenwich hour angle
GMT Greenwich Mean Time (Universal Time)
GP geographical position
Ha apparent altitude
Hc computed altitude
Ho observed altitude
HP horizontal parallax
Hs sextant altitude
Hz hertz (cycles per second)
IC index correction
kHz kilohertz
L latitude
L_1 latitude of departure
L_2 latitude of destination
Lv latitude of vertex
l difference in latitude
LAN local apparent noon
Lat. latitude
LHA local hour angle
LL lower limb
LMT local mean time
Lo longitude
Long. longitude
LOP line of position
Lo_1 longitude of departure
Lo_2 longitude of destination
Lov longitude of vertex
m meridional parts
MHz Megahertz

Min.	minute, minutes (time)
N	north
Na	nadir
No.	number
nm	nautical mile
NP	North pole
P	pole
Q	Polaris correction
R FIX	running fix
S	south, speed
Sec.	second, seconds (time)
SD	semidiameter
SHA	sidereal hour angle
sin	sine
$\sin^{-1}$	inverse sine
SP	South pole
T	toward (altitude difference), true (direction)
t	meridian angle
tan	tangent
$\tan^{-1}$	inverse tangent
UL	upper limb
UT	Universal Time (Greenwich Mean Time)
V	vertex
v	interpolation value, vertex
W	west
Z	azimuth angle (0° to 180° east-west), zenith
ZH	zenith height
Zn	true azimuth (0° to 360°)

APPENDIX B

EQUATIONS

INTERCEPT-AZIMUTH

Equation 1: $Hc = \sin^{-1}[(\sin L \times \sin d) + (\cos L \times \cos d \times \cos t)]$

Equation 2: $Z = \cos^{-1}\{[\sin d - (\sin L \times \sin Hc)]/(\cos L \times \cos Hc)\}$

LATITUDE AT LAN

Equation 3: $Hc = \sin^{-1}[\cos(L - d)]$

LATITUDE AT SUNSET

Equation 5: $t = \pm(Lo - GHA)$

Equation 9: $L = \tan^{-1}[-(\cos t/\tan d)]$

LONGITUDE BY TIME SIGHT

Equation 4: $t = \cos^{-1}\{[\sin Hc - (\sin L \times \sin d)]/(\cos L \times \cos d)\}$

Equation 5: $Lo = GHA \pm t$

LONGITUDE BY PRIME VERTICAL SIGHT

Equation 6: $Hc = \sin^{-1}(\sin d/\sin L)$

Equation 7: $t = \sin^{-1}(\cos Hc/\cos d)$

Equation 5: $Lo = GHA \pm t$

LONGITUDE BY TIMING SUNRISE OR SUNSET

Equation 8: $t = \cos^{-1}\{[\sin -.84° - (\sin L \times \sin d)]/(\cos L \times \cos d)\}$

Equation 9: $Lo = GHA \pm t$

SUNRISE AND SUNSET CALCULATIONS

Equation 8: $t = \cos^{-1}\{[\sin -.84 - (\sin L \times \sin d)]/(\cos L \times \cos d)\}$

Equation 10: $LMT\ (\text{mean sun}) = (180° \pm t)/15$

CIVIL TWILIGHT CALCULATION

Equation 11: $t = \cos^{-1}\{[\sin -6° - (\sin L \times \sin d)]/(\cos L \times \cos d)\}$

Equation 10: $LMT \text{ (mean sun)} = (180° \pm t)/15$

NAUTICAL TWILIGHT CALCULATION

Equation 12: $t = \cos^{-1}\{[\sin -12° - (\sin L \times \sin d)]/(\cos L \times \cos d)\}$

Equation 10: $LMT \text{ (mean sun)} = (180° \pm t)/15$

MOONRISE OR MOONSET CALCULATION

Equation 13: $t = \cos^{-1}\{[\sin(-.57 - SD + HP) - (\sin L \times \sin d)]/(\cos L \times \cos d)\}$

Equation 10: $LMT \text{ (mean moon)} = (180 \pm t)/15$

DIP

Equation 14: $Dip = -.97 \times \sqrt{\text{Ht. of Eye in ft.}}$

REFRACTION

Equation 15: $R = -.97 \times \tan\{Ha - \tan^{-1}[12 \times (Ha + 3)]\}$

GREAT CIRCLE SAILING

Equation 16: $D = \cos^{-1}[(\sin L_1 \times \sin L_2) + (\cos L_1 \times \cos L_2 \times \cos DLo)]$

Equation 17: $C_1 = \cos^{-1}\{[\sin L_2 - (\sin L_1 \times \cos D)]/(\cos L_1 \times \sin D)\}$

Equation 18: $Lv = \cos^{-1}(\cos L_1 \times \sin C_1$

Equation 19: $Dv_1 = \sin^{-1}[l/(\tan Lv \times \tan C_1)]$

Equation 20: $DLov = \cos^{-1}(\cos Dv_1 \times \sin C_1)$

Equation 21: $L_{1i} = \tan^{-1}(\cos Dv_1 \times \tan Lv)$

MERCATOR SAILING

Equation 22: $C = \tan^{-1}(DLo/m)$

Equation 23: $D = 1/\cos C$

APPENDIX C

GLOSSARY

ALTITUDE Angular elevation of a celestial object above the horizon.

APPARENT Appearing as real to the mind. Not necessarily factual.

APPARENT ALTITUDE (Ha) Altitude corrected for instrument error, personal error, and dip.

ARC DISTANCE Distance on a great circle between two points.

ARC MEASURE For navigational purposes, one minute of arc of latitude on the earth's surface is equal to one nautical mile (6076.1 ft.) and an angle of one minute at the center of the earth.

ARIES, FIRST POINT OF Point on the equinoctial where the sun's declination changes from South to North. The vernal equinox. The point on the equinoctial for locating the relative positions of stars (SHA).

ARTIFICIAL HORIZON A device used in lieu of an actual horizon when taking practice sextant sights ashore.

ASSUMED LATITUDE Latitude of an assumed position. Observer's latitude on the navigational triangle.

ASSUMED LONGITUDE Longitude of an assumed position. Observer's longitude on the navigational triangle.

ASSUMED POSITION (AP) Position of the observer on the navigational triangle, near to or at the DR position.

AZIMUTH ANGLE (Z) Interior angle on the navigational triangle between the observer's meridian and the vertical circle cutting through the body and the observer's zenith.

CARTESIAN COORDINATES Two coordinates at 90° used to locate a point like latitude and longitude on a Mercator chart. The (x, y) system.

CELESTIAL COORDINATES Coordinates used to define positions on the transparent surface of the celestial sphere.

CELESTIAL EQUATOR The great circle on the transparent surface of the celestial sphere generated by passing a plane through earth's equator and the celestial sphere. The equinoctial.

CELESTIAL HORIZON Great circle on the transparent surface of the celestial sphere generated by passing a plane at 90° to the zenith and nadir axis through the earth's center and the celestial sphere.

CELESTIAL NAVIGATION Navigation by observation of the positions of celestial bodies.

CELESTIAL POLES Extension of the terrestrial poles through the transparent surface of the celestial sphere.

CELESTIAL SPHERE Imaginary sphere of infinite radius with the earth as its center. All celestial bodies are imagined to be located on the transparent surface of the celestial sphere.

CELESTIAL TRIANGLE Spherical triangle on the surface of the celestial sphere formed by the observer's zenith, the celestial position of a body, and the elevated pole as vertices. Colatitude of the observer's zenith, codeclination of a body, and coaltitude of a body form the sides. Same as navigational triangle.

CHRONOMETER Instrument (clock) designed to keep accurate time.

CIRCLE OF EQUAL ALTITUDE All the points on the earth's surface where the altitude observations of a body would be identical at the same time. Also a circle of position.

CIVIL TWILIGHT Period when the sun is from 0° to 6° below the horizon.

COALTITUDE Complement of altitude (90° – altitude).

COAST PILOT Book containing localized information on coastal features, harbors, currents, weather, etc. for mariners.

COLATITUDE Complement of latitude (90° – latitude)

COMPASS ERROR Difference between true direction and compass heading.

COMPUTED ALTITUDE (Hc) Altitude computed with a calculator or extracted from tables for comparison with observed altitude to determine an LOP.

COURSE LINE LOP An LOP parallel with the true course sailed.

DEAD RECKONING The determination of position without celestial navigation. Using record of heading, speed, and distance.

DECLINATION (DEC. or d) Interior angle at the center of the celestial sphere between the equinoctial and a celestial body. Same as latitude of a celestial body.

DIP Angle between the true and apparent horizon.

ECLIPSE Partial or total obscuring of one celestial body by another.

ELEVATED POLE Polar axis North or South in the same hemisphere as the observer.

EQUATION OF TIME The difference in arc between the path of the true sun and the path of the fictitious mean sun expressed as time.

EQUINOCTIAL Celestial equator formed by the extension of the equitorial plane.

EQUINOX A point where the sun crosses the celestial equator.

FULL-HOUR MERIDIAN Meridian that is a multiple of 15°, east or west of the prime meridian, and is in the middle of a time zone, extending 7.5° on either side of the meridian.

GEOGRAPHIC POSITION (GP) Point on the earth's surface where a line from the center of the earth to a celestial body pierces the earth's surface. Altitude of the celestial body would be 90° at the GP.

GREAT CIRCLE Circle on the surface of a sphere formed by the intersection of the surface with a plane that passes through the center of the sphere. All meridians and the equator are great circles on the earth's surface.

GREAT CIRCLE COURSE Shortest distance between two points on the earth's surface.

GREAT CIRCLE SAILING Course, distance, and positions along a great circle course.

GREENWICH HOUR ANGLE (GHA) Angle between the prime meridian (Greenwich) and an hour circle of a celestial body measured in a westerly direction from 0° to 360°.

GREENWICH HOUR ANGLE OF ARIES Angle between the prime meridian (Greenwich) and the first point of Aries.

GREENWICH MEAN TIME (GMT) Local time at the prime meridian (0° longitude). Same as Coordinated Universal Time (UTC). Time of the mean sun as measured from the lower branch of the prime meridian (180° longitude).

GREENWICH MERIDIAN The prime meridian (0° longitude). Passes through the Royal Observatory, Greenwich, England. Reference standard for time and longitude.

HORIZON Plane at 90° to the zenith and nadir axis forming a great circle on the celestial sphere called the celestial horizon.

HORIZON, APPARENT Horizon as seen by an observer at sea. Differs from horizon by the dip for sun, stars, and planets. Differs from horizon by dip and HP for the moon.

HORIZON GLASS Glass attached to the frame of the sextant. Glass is half mirror and half see-through glass.

HORIZONTAL PARALLAX (HP) Maximum parallax when the moon is on the horizon.

HOUR CIRCLE Meridian analogous to longitude extended to the transparent surface of the celestial sphere.

INDEX ARM Part of the sextant that is moved by depressing the release lever.

INDEX ERROR Instrument error when index mirror is not aligned with the horizon.

INDEX MIRROR Mirror attached to the index arm of the sextant.

INFERIOR PLANETS Planets inside the earth's orbit around the sun. Venus, Mercury.

INSTRUMENT ERROR Constant error due to manufacture or adjustment of an instrument such as a sextant.

INTERCEPT Difference in minutes of arc between observed altitude (Ho) and computed altitude (Hc) of a celestial body and should not exceed 30′. The intercept is measured in nautical miles toward (T) if Ho is greater than Hc, and away (A) if Ho is less than Hc.

INTERNATIONAL DATE LINE Lower branch of the prime meridian (180°). Date increases one day (24 hrs.) crossing from W to E longitude. Date decreases one day (24 hrs.) crossing from E to W longitude.

INVERSE TRIGONOMETRIC FUNCTION Angle that is found from knowledge of the value of the function.

LATITUDE (LAT. or L) Angle made with the equator, center of the earth, and a designated point on the earth's surface. Measured N or S of the equator along the meridian of the point.

LAW OF COSINES FOR SPHERICAL TRIANGLES The cosine of the side of a spherical triangle is equal to the product of the cosines of the other two sides, plus the product of the sines of the other two sides and the cosine of their included angle: $\cos a = \cos b \cos c + \sin b \sin c \cos A$.

LAW OF SINES FOR SPHERICAL TRIANGLES The sine of any side of a spherical triangle divided by the sine of the angle opposite that side is equal to the sine of any other side divided by the sine of the angle opposite the other side: $\sin a/\sin A = \sin b/\sin B = \sin c/\sin C$.

LIMB Upper and lower points of tangency on the circumference of the sun or moon.

LINE OF POSITION (LOP) A line along which an observer is located at some point. In celestial navigation a small portion of a circle of position or a circle of equal altitude defining an infinite number of possible terrestrial positions of an observer.

LOCAL APPARENT NOON (LAN) The exact time that the sun is on the upper branch of the observer's meridian.

LOCAL HOUR ANGLE (LHA) Angle between the upper branch of the observer's meridian and an hour circle through celestial body, measured westerly from 0° to 360°.

LONGITUDE (Long. or Lo) Angle between the prime meridian and a point on a meridian on the earth's surface measured from the prime meridian E or W from 0° to 180° around the polar axis.

LOWER LIMB (LL) Point of tangency on the circumference of the sun or moon closest to the horizon.

LOWER TRANSIT Crossing of a celestial body across the lower branch of the observer's meridian.

MARCQ ST. HILAIRE METHOD Intercept-azimuth method of determining an LOP. Measured by the length of an intercept T or A from an AP in the direction Zn.

MEAN Average of any number of related observations. Found by dividing the sum of the related observations by the number of observations.

MEAN SUN Fictitious sun that completes every day, at a constant rate, in exactly 24 hours.

MEAN TIME Time measured with respect to the mean sun. Greenwich Mean Time or coordinated universal time is mean time.

MERCATOR CHART Chart constructed by projecting the earth's surface onto a cylinder from a point in the center of the earth. The equator is usually projected tangent with the cylindrical chart.

MERCATOR SAILING Computing course, distance, and points on a Mercator chart. Distance is measured as a function of latitude difference and course.

MERIDIAN Great circle created on the transparent surface of the celestial sphere or the surface of the earth (longitude) by passing a plane through the polar axis.

MERIDIAN ANGLE (t) Angle from the observer's meridian to the hour circle of a celestial body. Measured from 0° to 180°, E or W.

MERIDIAN PASSAGE Meridian transit.

MERIDIAN TRANSIT Exact time when a celestial body is on the observer's meridian.

MERIDIONAL PART (m) One minute of arc at the equator on a Mercator chart. Meridional parts are the number of minutes of arc required to compensate for the expanded latitude difference between two points on a Mercator chart.

MOONRISE Local time when the upper limb of the moon is first seen at the horizon.

MOONSET Local time when the upper limit of the moon is last seen at the horizon.

NADIR Point on the transparent surface of the celestial sphere in the opposite direction from the observer's zenith.

NAUTICAL ALMANAC Annual publication containing celestial data pertaining to celestial bodies and essential to the practice of celestial navigation. Data is arranged according to the seconds, minutes, hours, days, and months of a given year. Now includes tables for sight reduction.

NAUTICAL MILE One minute of arc on a great circle on the earth's surface. Measures 6,076 ft. or 1.15 statute miles.

NAUTICAL TWILIGHT Period when the sun is from 0° to 12° below the horizon.

NAVIGATIONAL PLANETS Venus, Mars, Jupiter, and Saturn. Data are provided in the *Nautical Almanac* for navigational planets.

NAVIGATIONAL TRIANGLE Spherical triangle defined by its elevated pole (NP) or (SP), zenith of an observer (Z), and geographical position of a celestial body (GP) as vertices and coaltitude, codeclination, and colatitude as sides.

OBSERVED ALTITUDE (Ho) Sextant altitude corrected for personal error, instrument error, dip, refraction, parallax (moon only), temperature, and pressure.

OBSERVER'S MERIDIAN Meridian on the transparent surface of the celestial sphere corresponding to the longitude of the observer.

PARALLACTIC ANGLE Angle of the navigational triangle between hour circle of the body and the vertical circle to the observer. Angle of least use to the navigator.

PARALLAX Angular difference between altitude measured at the surface of the earth and the earth's center.

PERSONAL ERROR Error incurred from observations of a specific observer.

PILOT CHART Charted information on wind strength and directions, currents, storm frequency and paths, temperatures, and other details of interest to mariners.

POLAR AXIS North and South poles forming axis of earth's rotation.

POLAR COORDINATES Two numbers used to locate a point on a plane. The first by its distance from the initial point, and the second by the angle the line makes with the reference line. Describes course and distance on a Mercator chart.

POLAR DISTANCE Angle between the elevated pole and a celestial body called codeclination.

POLAR-EQUATORIAL SYSTEM Celestial sphere defined by the polar axis and the celestial equator or equinoctial.

PRIME MERIDIAN Meridian passing through the Royal Observatory at Greenwich, England. Standard meridian for measurement of longitude and time. Longitude 0°.

PRIME VERTICAL CIRCLE Vertical circle at 90° or 270° (E or W) with respect to an observer.

PRIME VERTICAL SIGHT Sight timed such that a celestial body will be on the prime vertical circle and the resultant LOP will be the longitude.

RANDOM ERROR Non-systematic error due to unpredictable circumstances. Averages to zero in practice.

REFRACTION Angular change in direction of light beams as the result of passing through medium of changing density, such as the earth's atmosphere. The earth's atmosphere causes observations of altitude to appear higher than they actually are.

REVOLUTION Orbital displacment of a body (earth) around another (sun). Path can be circular, elliptical, etc.

RHUMB LINE Straight line on a Mercator chart cutting all meridians at the same angle.

ROTATION Circular movement of a celestial body around its own axis.

RUNNING FIX Terrestrial position determined by advancing LOPs to the time of the last LOP.

SAILING DIRECTIONS Published information for mariners for coastal piloting and navigation over various areas.

SAILINGS Method of determining the course to be followed to reach a predetermined destination. Includes determinations of course, distance, and waypoints.

SEMIDIAMETER (SD) Half the diameter of a celestial body such as the sun or moon; used to determine upper and lower limb corrections.

SEXTANT ALTITUDE (Hs) Altitude taken from the sextant after an observation and prior to applying any corrections.

SHADOW BEARING Bearing of the sun's shadow over the shadow pin of a magnetic compass ± 180°.

SHADOW PIN Vertical pin centered on the card of the magnetic compass for obscuring a small portion of the sun's rays so that a shadow bearing can be obtained.

SIDEREAL HOUR ANGLE (SHA) Angle between the first point of Aries and the hour circle of a star measured in a westerly direction.

SIGHT REDUCTION Determining an LOP from a celestial observation. Entails determination of intercept and azimuth from an AP.

SIGHT REDUCTION TABLES Computed solutions of the navigational triangle. Entering arguments are assumed latitude (Ass. Lat.), declination (True Dec.) and local hour angle (LHA) for obtaining computed altitude (Hc) and true azimuth (Zn).

SIMULTANEOUS FIX Terrestrial position obtained by observations of two or more celestial bodies at essentially the same time.

SMALL CIRCLE Circles described by passing a plane through a sphere without passing it through the center. Circles on the earth's surface describing latitude, except for the equator.

SPHERICAL TRIGONOMETRY Trigonometric equations used to

mathematically solve spherical triangle problems in sight reduction and the sailings.

STATUTE MILE Measure of distance in the United States. One statute mile is equal to 5,280ft. or 0.869 nm.

SUNRISE Local time when the upper limb of the sun is first seen at the horizon.

SUNSET Local time when the upper limb of the sun is last seen at the horizon.

SUPERIOR PLANETS Planets outside the earth's orbit around the sun. Mars, Jupiter, Saturn, Uranus, Neptune, Pluto.

SWINGING THE ARC Technique of rocking head and sextant in unison such that a celestial body will touch the horizon at the bottom of the described arc, ensuring that the sextant is perpendicular to the horizon.

TIME SIGHT Method of determining longitude when latitude is well known, by timing a sun sight approximately two hours after the sun's transit, computing t, GHA, and hence, longitude.

TRUE AZIMUTH (Zn) Bearing from the observer to a celestial body with respect to the North polar axis measured clockwise from 0° to 360°.

UNIVERSAL PLOTTING SHEET Plotting sheets constructed with fixed latitude separation and variable longitude separation to approximate portions of Mercator charts as needed for plotting dead reckoning and terrestial positions.

UPPER LIMB (UL) Point of tangency of the circumference of the sun or moon farthest from the horizon.

UPPER TRANSIT Crossing of a celestial body across the upper branch of the observer's meridian.

VERNAL EQUINOX The first point of Aries. Point on the equinoctial where the sun's declination changes from South to North.

VERTEX (V or v) Highest latitude attained by the actual or extended great circle course.

VERTICAL CIRCLE Great circle on the transparent surface of the celestial sphere forming an intersection with a celestial body, the observer's zenith (Z) and nadir (Na).

WORKING PERIOD Period between the end of civil and the end of nautical twilight following sunset, or the beginning of these periods at dawn. When the navigational celestial bodies and the horizon are optimally visible.

ZENITH (Z) Point on the transparent surface of the celestial sphere directly above the observer and extending through the center of the earth.

ZENITH HEIGHT (ZH) Angle between the observer's zenith and the observed altitude (Ho) of a celestial body.

ZENITH-HORIZON SYSTEM Celestial sphere defined by the zenith (Z) and nadir (Na) of the observer as poles and the horizontal plane at 90° to the zenith and nadir through the middle of the earth and extending to the transparent surface of the celestial sphere to form the celestial horizon.

ZONE TIME Greenwich mean time ± one hour difference every 15° of longitude. Meridian at each 15° of longitude is called a full-hour meridian. A zone is 15° wide and extends 7.5° on either side of the full-hour meridian. Time zones on land usually have boundaries that are very irregular and arbitrary.

APPENDIX D

WORKSHEET

LINE OF POSITION WORKSHEET

NO. ________ AREA ________

BODY	DATE	D. R. LAT.	D. R. LONG.	LOG	WWV	STOP WATCH	LOCAL TIME

TIME	HR.	MIN.	SEC.
WWV			
Stop Watch			
GMT			

SUN, STARS, & PLANETS SEXTANT (ALMANAC)

Hs	Ha
IC	Corr.
Dip	Add'l Corr.
Ha	Ho

MOON ONLY SEXTANT (ALMANAC)

Corr.		Hs
Add'l Corr.		IC
HP	Corr.	Dip
Total		Ha
		Total
		Ho

ASS. POSITION LHA (ALMANAC)

GHA Hours	
Min. & Sec.	
Stars Only SHA	
Moon & Planets Only V:	Corr.
360°	
Total GHA	
Ass. Long.	
LHA	

INTERCEPT AZIMUTH – FIRST ENTRY

Lat.	LHA	A	A°	A′
		B $\pm$	$Z_1 \pm$	
		Dec. $\pm$		
Sum: B+Dec.		F	F°	F′

PLOTTING

Ass. Long.		
Zn	Ass. Lat.	
Intercept	T	A

INTERCEPT AZIMUTH – SECOND ENTRY

A°	F°	H	P°	$Z_2 \pm$
F′	P°	$Corr._1 \pm$	Z_2°	
	Sum		Z_1	
A′	Z_2°	$Corr._2 \pm$	Z_2	
	True Hc		Z	
	True Ho		Zn	
	Intercept		T	A

DECLINATION (ALMANAC)

Dec. Hours	
"d" $\pm$	Corr. $\pm$
True Dec.	

LINE OF POSITION WORKSHEET

NO. ______ AREA ______

BODY	DATE	D. R. LAT.	D. R. LONG.	LOG	WWV	STOP WATCH	LOCAL TIME

TIME

	HR.	MIN.	SEC.
WWV			
Stop Watch			
GMT			

ASS. POSITION LHA (ALMANAC)

GHA Hours	
Min. & Sec.	
Stars Only SHA	
Moon & Planets Only V:	Corr.
360°	
Total GHA	
Ass. Long.	
LHA	

DECLINATION (ALMANAC)

Dec. Hours	
"d" $\pm$	Corr. $\pm$
True Dec.	

SUN, STARS, & PLANETS SEXTANT (ALMANAC)

Hs	Ha
IC	Corr.
Dip	Add'l Corr.
Ha	Ho

MOON ONLY SEXTANT (ALMANAC)

Corr.		Hs
Add'l Corr.		IC
HP	Corr.	Dip
Total		Ha
	Total	
	Ho	

INTERCEPT AZIMUTH – FIRST ENTRY

Lat.	LHA	A	A°	A′
		B $\pm$	Z_1 $\pm$	
		Dec. $\pm$		
	Sum: B+Dec.	F	F°	F′

INTERCEPT AZIMUTH – SECOND ENTRY

A°	F°	H	P°	Z_2 $\pm$
F′	P°	$Corr._1$ $\pm$	Z_2°	
	Sum		Z_1	
A′	Z_2°	$Corr._2$ $\pm$	Z_2	
	True Hc		Z	
	True Ho		Zn	
	Intercept		T	A

PLOTTING

Ass. Long.		
Zn	Ass. Lat.	
Intercept	T	A

LINE OF POSITION WORKSHEET

NO. ______ AREA ______

BODY	DATE	D. R. LAT.	D. R. LONG.	LOG	WWV	STOP WATCH	LOCAL TIME

TIME	HR.	MIN.	SEC.
WWV			
Stop Watch			
GMT			

SUN, STARS, & PLANETS SEXTANT (ALMANAC)

Hs	Ha
IC	Corr.
Dip	Add'l Corr.
Ha	Ho

MOON ONLY SEXTANT (ALMANAC)

Corr.		Hs
Add'l Corr.		IC
HP	Corr.	Dip
Total		Ha
	Total	
	Ho	

ASS. POSITION LHA (ALMANAC)

GHA Hours	
Min. & Sec.	
Stars Only SHA	
Moon & Planets Only V:	Corr.
360°	
Total GHA	
Ass. Long.	
LHA	

INTERCEPT AZIMUTH – FIRST ENTRY

Lat.	LHA	A	A°	A′
		B $\pm$	$Z_1 \pm$	
		Dec. $\pm$		
	Sum: B+Dec.	F	F°	F′

PLOTTING

Ass. Long.		
Zn	Ass. Lat.	
Intercept	T	A

INTERCEPT AZIMUTH – SECOND ENTRY

A°	F°	H	P°	$Z_2 \pm$
F′	P°	$Corr._1 \pm$	Z_2°	
	Sum		Z_1	
A′	Z_2°	$Corr._2 \pm$	Z_2	
	True Hc		Z	
	True Ho		Zn	
	Intercept		T	A

DECLINATION (ALMANAC)

Dec. Hours	
"d" $\pm$	Corr. $\pm$
True Dec.	

LINE OF POSITION WORKSHEET

NO. ______ AREA ______

BODY	DATE	D. R. LAT.	D. R. LONG.	LOG	WWV	STOP WATCH	LOCAL TIME

TIME	HR.	MIN.	SEC.
WWV			
Stop Watch			
GMT			

SUN, STARS, & PLANETS SEXTANT (ALMANAC)

Hs	Ha
IC	Corr.
Dip	Add'l Corr.
Ha	Ho

MOON ONLY SEXTANT (ALMANAC)

Corr.		Hs
Add'l Corr.		IC
HP	Corr.	Dip
Total		Ha
	Total	
	Ho	

ASS. POSITION LHA (ALMANAC)

GHA Hours	
Min. & Sec.	
Stars Only SHA	
Moon & Planets Only V:	Corr.
360°	
Total GHA	
Ass. Long.	
LHA	

INTERCEPT AZIMUTH – FIRST ENTRY

Lat.	LHA	A	A°	A'
		B $\pm$	$Z_1 \pm$	
		Dec. $\pm$		
Sum: B+Dec.		F	F°	F'

PLOTTING

Ass. Long.		
Zn	Ass. Lat.	
Intercept	T	A

DECLINATION (ALMANAC)

Dec. Hours	
"d" $\pm$	Corr. $\pm$
True Dec.	

INTERCEPT AZIMUTH – SECOND ENTRY

A°	F°	H	P°	$Z_2 \pm$
F'	P°	$Corr._1 \pm$	Z_2°	
	Sum		Z_1	
A'	Z_2°	$Corr._2 \pm$	Z_2	
	True Hc		Z	
	True Ho		Zn	
	Intercept		T	A

APPENDIX E

REFERENCES

Bowditch, Nathaniel, *American Practical Navigator*, Washington, D.C.: U.S. Government Printing Office, Vol. I 1984, Vol. II 1981.

Duttons, *Navigation and Piloting*, Edited by Maloney, Albert S. 13th ed. Annapolis, Md.: Naval Institute Press, 1978.

Kittredge, Robert Y., *Self-Taught Navigation*, 2nd ed. Flagstaff, Az.: Northland Press, 1973.

Nautical Almanac, Washington, D.C.: U.S. Printing Government Printing Office, annually.

Ocean Passages for the World, Prepared by Jenkins, H.L. 3rd ed. Great Britain: Hydrographer of the Navy, 1973.

Page, Raymond, "Computerized Celestial Navigation." *Sea*, May 1983, page 42.

Sight Reduction Tables for Air Navigation, Pub. No. 249, 3 vols. Washington, D.C.: U.S. Government Printing Office, 1985.

Turner, Merle B., *Celestial for the Cruising Navigator*, 1st ed. Centreville, Md.: Cornell Maritime Press, 1986.